A SHEARWATER BOOK

The Value of Life

Stephen R. Kellert

The Value of Life

Biological Diversity and Human Society

ISLAND PRESS / Shearwater Books
Washington, D.C. • Covelo, California

A Shearwater Book
published by Island Press

Library of Congress Cataloging-in-Publication Data
Kellert, Stephen R.
The value of life : biological diversity and human society/
Stephen R. Kellert.
p. cm.
Includes biographical references (p.) and index.
ISBN 1-55963-317-4 (cloth)
1. Human ecology—Philosophy. 2. Philosophy of nature. 3. Environmental degradation—Moral and ethical aspects. 4. Nature conservation—Philosophy. 5. Biological diversity conservation—Philosophy. I. Title.
GF21.K47 1996
179′.1—dc20 95-32210
CIP

Printed on recycled, acid-free paper

Manufactured in the United States of America
10 9 8 7 6 5 4 3 2

To Cilla, Emily, Libby,
and the boys
for all their
love and inspiration.

CONTENTS

List of Tables and Figures xi

Acknowledgments xiii

Prologue xv

PART ONE: Universals

Chapter 1: Introduction 3

Chapter 2: Values 9

PART TWO: Variations

Chapter 3: American Society 37

Chapter 4: Activities 64

Chapter 5: Species 99

Chapter 6: Culture 131

PART THREE: Applications

Chapter 7: Endangered Species 155

Chapter 8: Conserving Biological Diversity 185

Chapter 9: Education and Ethics 209

Notes 219

Index 249

TABLES AND FIGURES

TABLES

1. Typology of Basic Values *38*
2. Japanese and American Values *138*

FIGURES

1. American Mean Attitude and Knowledge Scores *41*
2. Utilitarian Value: 1900–1976 *43*
3. Ecologistic Value: 1900–1976 *45*
4. Attitudes That Decrease with Age: Children's Negativistic, Humanistic, and Dominionistic Values *48*
5. Attitudes That Increase with Age: Children's Ecologistic, Moralistic, and Naturalistic Values *48*
6. Gender Value Differences *52*
7. Education Value Differences Among Adults *55*
8. Utilitarian Urban/Rural Value Differences *58*
9. Moralistic Urban/Rural Value Differences *58*

10. Ethnic Value Differences 61
11. American Attitudes Toward Hunting 71
12. Values Among Birders 83
13. Childhood Cruelty Toward Animals 95
14. Factors Shaping Attitudes Toward Wildlife 100
15. Attitudes Toward Wolves in Minnesota 109
16. Utilitarian Attitudes Toward Marine Mammals in Canada 117
17. Moralistic Attitudes Toward Marine Mammals in Canada 117
18. Attitudes Toward Whaling 119
19. Attitudes Toward Invertebrates 125
20. Wildlife Policy Framework 156
21. Attitudes Toward Water Projects That Endanger Fish Species 165
22. Attitudes Toward Protecting Endangered Species 170

ACKNOWLEDGMENTS

This book presents the results and ideas of various projects I have been involved with for nearly twenty years. Many people played indispensable roles in these efforts. Since nearly all of them are mentioned in the text or referenced in the chapter notes, I will not burden the reader by citing these persons again here. I do want, however, to express my collective gratitude to all of them for having so ably assisted me in the development of my interests, ideas, and ethics. I hope this book does justice to their many contributions toward creating a more decent, affirming, and enriching relationship between people and the diversity of life.

I want also to express my particular appreciation to Barbara Dean of Island Press whose encouragement, professional advice, and friendship have been so critical in writing this book.

Finally, I wish to extend my special thanks to Scott McVay and the Geraldine R. Dodge Foundation for their wonderfully generous intellectual and financial support which has made possible so many of the conclusions reported in this book.

PROLOGUE

I struggle, like many in our society, with the need to fashion a coherent sense of community and connection out of fragments. Often I feel sliced into separate, seemingly incompatible roles, each marked by complexity and abstraction, of relevance to only a small audience of others. Few opportunities exist for tying these many threads of a life together—for uniting my professional identity with the traditional roles of parent, spouse, friend, citizen, community member, and participant in a natural, less human-built and dominated world. The historical link between work, community, and nature, once the basis for a secure and sustainable sense of place, has eroded. A new kind of alienation has taken its stead.

I find myself wrestling with such demons this early November morning, as my disconnected professional and personal roles clash with my desire for a more cohesive existence. Outside I hear the neighborhood stir with preparations for the new day, the busy pulse of work, school, and commerce projecting a hum of energy and purpose. What seems lacking in all this activity, though, is a sence of integrity—an integration of work and community, a harmony of place and environment.

I choose a time-honored tradition to make the transition from my current disconnection to the glimpse of a greater unity of direction and purpose. I gather my dogs for a short walk to the nearby park and its meandering river. The thick-set, hang-jowled, clumber spaniel and the sprightly poodle-like mixed breed anticipate my intentions, straining with enthusiasm, their keen senses poised in expectation of our ritual seeking of forests and watercourses.

Exiting the house we gather rhythm and poise as we spy a quarter-mile away, the nearly thousand acres of undeveloped oasis. The park's towering feature, a huge traprock ridge, looms high above, perpendicular cliffs of red rock signifying the hardened lava which has survived the millennia of eroding years that leveled the soft surrounding sandstone. We aim for the floodplain, where the river snakes along the base of the great columnar cliffs. We cross a busy intersection before entering the bottomland forest. As the traffic speeds past, little chance exists that the drivers and I will recognize our basic commonality.

The dogs strain with excitement as we descend the path to the river. We progress a few hundred yards and already the roar of the rushing traffic seems swallowed by the vegetative mass of the thick forest. A jostling of middle-aged oaks, tulips, maples, hickories, ashes, beeches, sassafras, willows, locusts, pines, hemlocks, dogwoods, and laurels has digested the muffled roar like some monstrous appetite, the machine sounds displaced by a great ambient thud. We have entered another world: richly textured interlacings of living matter, sweet soil, flowing waters, sand and stone, altogether calm and reassuring.

We proceed along the trail, repeatedly distracted by details as we pause to examine and explore, feeding our appetite for discovery, although we have followed this path many times before. We maneuver through the deciduous forest that soon levels onto the floodplain. Life crowds all around us—a fever of animation reflected in a profusion of songbirds, small mammals, insects, trees, bushes, rustling leaves, and more. We have become enchanted by a kaleidoscope of living abundance and diversity. I use my eyes, mostly, sometimes my ears, the dogs utterly lost in an exotic world of richly textured scents and smells.

As the November trees have lost most of their leaves, the shallow, slow-moving river soon comes into view. A kingfisher's distinctive alarm sounds from a nearby bank. The path winds through a canopy of spicebush, bittersweet, wild grape, and creeper, heavy with autumn seeds and berries. The rustling of resident and migrating songbirds hints at the gathering of winter fat. A flock of migrating robins engorges itself nearby. Small groups of mallards, geese, and swans pocket corners of the meandering river. A trellis of bluish-black, red, and yellow berries forms a graceful arch over the path, a welcoming embrace, joining the adjacent willows and locusts.

Winding through a belt of marsh grass and cattails, the path eventually leads to a footbridge elegantly spanning the river. It is a human creation distinctively in harmony with its landscape. I pause on top of the bridge. Below me I catch sight of the ancient elegance of a great blue heron crouching close to the river's edge, the cattails and phragmites stirring in the frosty morning air hinting at the coming of winter cold. Willows hang beside the riverbank casting pale green and yellow reflections on the slow-moving surface. Straining to see into the murky waters, I barely discern the ghostlike shadows of passing bass and shiners, two among the many fish who commingle in the brackish waters influenced by the nearby saltwater sound. The reflection of an undulating flicker catches my eye, its yellow glint and colorful underbody silhouetted against the river's surface.

The stress of disconnected realities, the uncertainty of place and relation in an age of confusion, flow from my shoulders like sap from a wounded tree, the tension absorbed by the soft, forgiving ground. I feel settled. Not just a sense of relaxation, but an approaching tranquility. I experience the promise of well-being flowing from a feeling of connection with the varied life and nonlife around me. I feel an affinity with this vibrant landscape set against a backdrop of contemporary sameness and artificiality. And, there is more. A web of relationships links me with this pocket of nature, some physical, some emotional, a few intellectual, even a flirtation with the spiritual. Intimate affiliation with living diversity offers me knowledge and kinship, and I am nourished by the association.

I take pleasure from the red and yellow vines and deep blue berries of the bittersweet canopy, the thick willows and locusts edging the river, the huge basaltic rock hovering above. There is satisfaction, too, in witnessing the fullness of the trees and the clouds reflected in the slow river. A raft of brants and mallards, their finely contrasting colors, represents yet another attraction: their unexpected burst into flight expresses spontaneity springing into motion.

The successional forest, the wetlands, the surrounding vegetation, the myriad of biotic and abiotic relations—all feed the impression of ecological connection among the many parts, a systematic alliance that both transcends and includes my presence. Apart from the intellectual insight, I feel charged by my role as member of the ecological enterprise.

The physical exertion, even in this tamed and diminished wildness, reaffirms my ancient roots and spurs confidence in my capacity for curiosity, exploration, and skill. My imagination aroused, I seek experience and understanding by pushing deeper into nature's maze. The magic well has again worked its curious transformation: the more I search, the more I recognize how much deeper and perhaps unending the searching might be.[1]

I gather comfort from a material dependency, a gentle and sustainable utilization expressed along the way. I have encountered compatible instances of practical human intervention—the gentle arch of the footbridge, the old stone dam still impounding a portion of the city's water supply, the remains of a gun factory signifying the initial stirrings of an industrial revolution. Most of all, the still healthy river circulates the city's nutrients, controls its floods, decomposes its wastes, offers a nursery for its commercial fish, provides a host of free environmental services upon which all life, mine included, depends.

I take sustenance, emotional and spiritual, from an ineffable feeling of kinship with the many creatures in this oasis of urban nature. I feel bonded with the dogs, of course, but also with the waterfowl, songbirds, chipmunks, even the invertebrates. Despite all the variety and diversity, I am comforted, and inspired, by the knowledge of an extraordinary degree of shared molecular and genetic relation. Even in this modern context of concrete, steel, and glass, there persists more life, of which I am part, than in all the dead stars and planets of the vast universe as we know it.

The sum of these affiliations with the living diversity which surrounds me translates into a sense of wholeness, a reminder of an underlying order, perhaps even purpose. A new reassurance has muted my earlier anxieties. My brief visit to the magic well has readied me for the tasks and challenges of the day. I feel invigorated intellectually, engaged emotionally, enlivened aesthetically, assured spiritually. My respite from the modern temper, and its sometimes overwhelming isolation, has allowed a timeless connection to emerge. As the dogs and I leave the park, an ancient Ojibway expression comes to mind: "Sometimes I go about pitying myself, and all the time I am being carried on great winds across the sky."[2]

Part One

Universals

CHAPTER 1

Introduction

THIS BOOK is about the value of living diversity—how these values are integral to what it means to be fully human, yet how they are increasingly threatened by a massive hemorrhaging of life on earth. Although the connection between these issues has become clearer to me of late, this recognition has emerged only after years of researching how people value living diversity: emotionally, intellectually, and materially.

I first became interested in the issue of how people value nature and wildlife two decades ago. From my innocent perch of the time, I was primarily concerned with the problem of how the effective management of wildlife often seemed less a problem of manipulating animals and their habitats than managing our own species' often callous and destructive disregard for much of the natural world. This perspective had certainly not originated with me. Aldo Leopold, one of the wildlife profession's pioneering ecologists, had suggested more than a half century before: "The problem of [wildlife] is not how we shall handle the [animals]. . . . The real problem is . . . human management. Wildlife management is comparatively easy; human management difficult."[1] I was struck, nonetheless, by how little systematic research had been done over the intervening years to explore the human/animal/nature relationship.[2]

Fortunately, during this early stage in my career, the U.S. Fish and Wildlife Service (FWS) also became interested in American values and perceptions of wildlife and its conservation. The FWS seemed motivated by increasing concern about what appeared to be new trends in American relationships to wildlife—particularly attitudes that challenged many of the service's traditional emphases on managing wildlife uses, mainly sport hunting and fishing. The FWS also appeared uncertain about new regulatory responsibilities imposed by the passage of the Endangered Species Act and Marine Mammal Protection Act in the early 1970s. They additionally wondered what was the motivation behind the explosion of wildlife recreational interest, particularly activities like birding, wildlife viewing, ecotourism, and others. All these changes represented significant new management challenges at the time, and the FWS believed an investigation of American values and behavior toward wildlife might better equip it for dealing with these profound shifts in American society.

A little historical perspective on the Fish and Wildlife Service is needed here to clarify its interest in this research. The service had traditionally focused on managing sport hunting and fishing and the hundred or more deer, ducks, salmonids, and other game species associated with these activities. This emphasis had served the profession well. Indeed, it was the basis for the generation of dependable revenues through the taxing and licensing of sportsmen, the elimination of the commercial wildlife trade, helping to restore depleted game species, and promoting the development of the wildlife management field. But it had also led to a strong financial, political, and ideological dependence on sport hunters and fishers. The price of this reliance had become a narrow management focus—and the exclusion of most of the public from the wildlife profession's inner circles.[3]

Two trends had eroded this tight logic by the 1970s. The first trend involved two linked phenomena: growing opposition to sport hunting and growing interest in nonconsumptive wildlife activities. Most wildlife enthusiasts do not object to hunting, but all antihunters are by definition nonconsumptive users and thus the two groups became lumped together in the minds of many sportsmen and wildlife managers. The Fish and Wildlife Service hoped a national study might discover the motivation behind these two presumably connected phenomena.

The second trend was the passage of major new laws at considerable variance with the wildlife management profession's traditional approach. The Endangered Species Act of 1973, the Marine Mammal Protection Act of 1972, the Convention on International Trade in Endangered Species of 1973, the National Environmental Policy Act of 1969, and other environmental legislation—all suggested a widening recognition of the need to protect natural systems, biological diversity, and rare and endangered species. These new requirements meant vastly expanded and unprecedented regulatory responsibilities for the Fish and Wildlife Service, an agency that had restricted itself largely to providing services for a fairly narrow clientele. These dramatic legislative changes created considerable administrative uncertainty for this traditionally low-profile agency—as well as political risks. The FWS worried about public support for these new regulatory responsibilities, particularly when they conflicted with powerful political and economic interests. Few recognized at the time how the problem of endangered species would, in just two decades, erupt into a firestorm of crisis proportions involving a projected loss of tens of thousands of creatures and a scale of biological impoverishment unprecedented in human history.

The Fish and Wildlife Service believed that by discovering the wildlife-related values, interests, and activities of the American public through a national study, it might be possible to manage wildlife in a more socially acceptable manner.[4] My study had several goals: the more equitable allocation of resources among users, a better basis for mitigating conflicts among wildlife interest groups, ascertaining support for protecting and restoring rare and endangered wildlife, educating the public about the value of wildlife and its conservation, and more fully understanding trends in American perceptions and uses of animals and the natural environment. The study's most significant challenge, given the extraordinary diversity of feelings and beliefs people direct at the natural world, was developing a means for classifying and measuring people's values of wildlife and nature. Logic suggested and science dictated the possibility of devising a set of basic values toward animals and the natural environment that might be systematically, empirically, and quantitatively studied across varying groups in society.

The first step was to delineate a taxonomy of basic values as a way of organizing and describing people's feelings and beliefs about animals and nature.[5] This conceptual framework is described in detail in the following chapter, but here it might help to note the nine essential values: an *aesthetic* attraction for animals and nature, a *dominionistic* interest in exercising mastery and control over wildlife, an *ecologistic and scientific* inclination to understand the biological functioning of organisms and their habitat, a *humanistic* affection and emotional bonding with animals, a *moralistic* concern for ethical relations with the natural world, a *naturalistic* interest in experiencing direct contact with wildlife and the outdoors, a *symbolic* use of animals and nature for communication and thought, a *utilitarian* interest in pragmatically exploiting wildlife and nature, and a *negativistic* avoidance of animals and the natural environment for reasons of fear, dislike, or indifference.

The development of this typology of basic values facilitated the measurement of the American public's attitudes toward wildlife and its conservation. During the course of this research and many subsequent studies since, it became more and more apparent that these patterns of thought might reflect universal dispositions toward nature somewhat independent of group affiliation, history, and culture. Indeed, the ubiquitous expression of the values suggested they might constitute basic tendencies—tendencies rooted in the biological character of the human species despite the molding and shaping influence of learning and experience.

As these views developed, I encountered the work of the American biologist Edward O. Wilson, particularly his writings on sociobiology and the concept of biophilia. Wilson defines biophilia as people's "innate tendency to focus on life and lifelike processes."[6] Could it be that my typology of basic values reflected a physical, emotional, and intellectual tendency among humans to affiliate with nature and living diversity? The values may, in other words, have developed during the long course of evolutionary time because of their functional significance. A subsequent book edited by Wilson and myself, *The Biophilia Hypothesis,* further elaborated this human affinity for life shaped by the formative influence of experience, learning, and culture.[7]

Many implications stem from the notion that people have a fundamental physical, emotional, and intellectual dependence on nature and living diversity. Above all, the meaningful and satisfying experience of these values may represent a vital expression of healthy human functioning and relationship to the natural world. Conversely, the erosion of this dependence on nature might signify considerable risk to humans materially, affectively, cognitively, and even spiritually. Most discussions of the harmful impacts of the species extinctions occurring annually—currently estimated at 15,000 to 30,000—have focused on the loss of material benefits to people such as fewer medicines, agricultural products, or diminished ecosystem functioning.[8] These losses certainly represent substantial threats to human well-being, but the biophilia notion suggests that far more may be at stake than just the diminution of people's material options. The degradation of life on earth might also signify the possibility of diminished emotional and intellectual well-being and capacity.

This book delineates these basic values of living diversity, their presumed importance to the realization of human functioning, and the threat posed by the current biodiversity crisis to our species' physical, emotional, and intellectual experience. Chapter 2 offers a detailed description of the nine basic values of animals and nature and connects these perceptions to human evolutionary development. The current large-scale loss of biological diversity is described, as well, particularly its possible impact on our fundamental dependence on nature and wildlife.

Although these basic values are depicted as inborn biological tendencies, they are greatly influenced by learning, culture, and experience. Part Two of the book considers the modifying effect on the content and expression of these values exerted by human demography, activity, relationship to varying species, and culture. Chapter 3 examines value differences in American society, particularly among different age, gender, education, occupation, urban/rural, and ethnic groups. Chapter 4 explores the influence of diverse animal-related experiences on perceptions of nature and wildlife including hunting, birding, zoos, television and film viewing, and abusing animals. The effect of diverse species on the human psyche is examined in Chapter 5, illustrated by attitudes toward wolves, whales, and invertebrates, insects in particular. Part Two concludes by assessing the role of culture in Chapter 6, especially value differences among Eastern and Western societies, the world's great industrial superpowers (the United States, Japan, and Germany), and views among developing non-Western nations, illustrated by Botswana.

The book's final section, Part Three, considers the application of understanding human values of living diversity in a variety of policy and management contexts. The complex problem of endangered species protection is examined in Chapter 7. Chapter 8 explores the general challenge of conserving biological diversity—particularly human competition and exploitation of biological resources in both rural and urban settings—and the need to develop more effective wildlife management institutions and structures.

The book's final chapter focuses on the indispensable role of education and ethics if we are to reduce the current hemorrhaging of life on earth. This chapter returns to the initial consideration of how people depend on a vast complex of subtle relationships with nature and living diversity to achieve lives rich in meaning and value. Modern society has embraced a dangerous illusion in coming to believe it can live apart from nature. Our ethical and institutional structures must acknowledge instead how much human life depends on healthy relationships with nature and living diversity. We need to relearn Henry Beston's suggestion of a half century ago:

> Whatever attitude to human existence you fashion for yourself, know that it is valid only if it be the shadow of an attitude to Nature. A human life, so often likened to a spectacle upon a stage, is more justly a ritual. The ancient values of dignity, beauty, and poetry which sustain it are of Nature's inspiration; they are born of the mystery and beauty of the world. Do no dishonor to the earth lest you dishonor the spirit of man.[9]

CHAPTER 2

Values

THIS CHAPTER describes the basic values of nature and explains their adaptational significance in human development. Although these values are rooted in human biology, as noted earlier, they are shaped by the formative influence of experience, learning, and culture. Indeed, the values may be expressed in diverse ways, and in the next several chapters, we will examine the shaping influence of demography, activity, culture, and species on people's basic perceptions of nature and living diversity.

Our human identity and fulfillment depend to a great extent on the satisfactory expression of these values of living diversity. The notion of biophilia, as we shall see, suggests that each of these values reflects a profound human craving for affiliating with nature and wildlife. The erosion or dysfunctional expression of these values can lead to a deprived and diminished existence. Without society's support and reinforcement, the values may manifest themselves marginally, but as elements of human biology they will remain frustrated. This may explain our current vicious cycle: society's denial of the importance of a rich and rewarding relationship with nature contributes to the extinction crisis, which, in turn, further alienates people from the natural world.

Nine Basic Values

The relationship between the values, the notion of biophilia, and the current large-scale loss of biological diversity will be discussed later in the chapter. But before turning to this complex relationship, the nine basic values of nature and living diversity need to be described. Each value is given a descriptive name in the following order of presentation: utilitarian, naturalistic, ecologistic-scientific, aesthetic, symbolic, dominionistic, humanistic, moralistic, and negativistic. These terms are just labels of convenience, however, not terminological straitjackets. And although their progression reflects a certain narrative logic, it is not meant to indicate their order of importance.

UTILITARIAN

The utilitarian value emphasizes the many ways humans derive material benefit from the diversity of life. The term "utilitarian" represents something of a misnomer, however, as all the values have utility insofar as they reflect some benefit to people. The conventional idea of utilitarian used here reflects the traditional notion of material benefit derived from exploiting nature to satisfy various human needs and desires.

Many plant and animal species provide material benefits to people in the form of food, medicine, clothing, tools, and other products. Most people recognize this dependence in nonindustrial societies, particularly among preliterate tribal hunter-gatherers, pastoralists, and others. Yet many developing nations still derive most of their output from extracting and exploiting wild living resources. Even industrially advanced countries such as Japan secure much of their food from exploiting wild fish stocks, and nearly 5 percent of the American economy has been found to derive from utilizing wild living species.[1]

Recent years have seen an expanded appreciation of the utilitarian value of nature and living diversity—particularly the future benefits that might be obtained from exploiting the genetic, biochemical, and physical properties of plant and animal species, many of them still insufficiently studied. We are beginning to recognize, too, the undiscovered significance of various obscure and unknown species. Only a small fraction of the many plants containing alkaloids, for example, an organic compound possessing anticancer properties, have been tested for their possible medicinal use. It has been estimated that some 25 to 40 percent of the world's current pharmaceutical products originated in a wild plant or animal species, and much of today's agricultural production depends on genetic improvement by a dwindling reservoir of wild plants.

Take, for example, two recent illustrations of this medical and agricultural dependence on wild living diversity: drugs derived from a single tropical plant, the rosy periwinkle, used to treat blood-related cancers, and a wild corn important in developing a new strain of domestic corn resistant to blight. Both the wild relative of agricultural corn and the rosy periwinkle nearly became extinct due to deforestation in, respectively, Mexico and Madagascar. These species represent but a small fraction of the many actual and potential medical, agricultural, industrial, and other products people obtain from wild living resources—and rapid advances in molecular biology, genetics, and bioengineering make this exploitation increasingly possible. This expanded utilitarian value of biological diversity suggests the folly of exterminating species just to satisfy short-term and unsustainable demands for timber, wildlife, minerals, and other products.

Beyond these benefits to society at large, people often obtain great satisfaction from their personal utilitarian experience of nature and living diversity. There is obvious benefit in picking berries, chopping firewood, harvesting wild animals, training dogs, and so on. But an intrinsic pleasure can also be derived from this participation in the movement of energy and material through varying cycles of life. No matter how mechanized and removed industrial society becomes from natural processes, there remains for many people a compelling need to feel connected to the practical utilization of nature and living diversity.[2]

NATURALISTIC

The naturalistic value emphasizes the many satisfactions people obtain from the direct experience of nature and wildlife. This value reflects the pleasure we get from exploring and discovering nature's complexity and variety. Indeed, the satisfactions people derive from contact with living diversity may be among the most ancient pleasures obtained from interacting with the natural world—particularly the more vivid plants and animals.[3]

Today the naturalistic experience often takes expression through formally organized recreation: birding, fishing, hunting, whalewatching, wildlife tourism, visiting zoos, and the like. People also derive naturalistic satisfaction from wandering the various woods, prairies, beaches, wetlands, and other natural areas. Living diversity is still an unrivaled context for engaging the human spirit of curiosity, exploration, and discovery, in an almost childlike manner, independent of age. A sense of permanence, simplicity, and pleasure often stems from experiencing unspoiled nature, directly observing wildlife, and participating in ancient rhythms.

Various studies have documented the many rewards of the naturalistic experience, among them relaxation, calm, and peace of mind.[4] Additional benefits may include enhanced intellectual growth, creativity, and imagination. As Seilstad suggests: "The surest way to enrich the knowledge pool that will keep the flywheel of cultural evolution turning is to nourish the human spirit of curiosity." Certainly immersion in nature can heighten a sense of vividness and widen the opportunity for discovery. These physical, emotional, and intellectual benefits have been revealed in studies of the outdoor recreation experience—for example, investigations have noted diminished tension, greater peace of mind, and an enhanced capacity for creativity from observing and discovering diversity in nature. Summarizing this research, Roger Ulrich concludes: "A consistent finding in well over 100 studies of recreation experiences in wilderness and urban nature areas has been that stress mitigation is one of the most important verbally expressed and perceived benefits."[5]

The naturalistic experience can also sharpen one's sensitivity to detail as the senses become more attuned to the moment—instilling a sense of living in time rather than passing through it. Moreover, a sharpened vitality and awareness can derive from a profound involvement in nature.

Intellectual stimulation, physical fitness, enhanced creativity—all may result from these encounters with the natural world.

ECOLOGISTIC-SCIENTIFIC

Although both the scientific and the ecologistic perspectives reflect an emphasis on biophysical patterns, structures, and functions of nature, there are important differences. The ecologistic view is a more integrative approach to the natural world emphasizing interdependence among species and natural habitats. The scientific understanding of nature, by contrast, tends to stress structures and processes below the level of whole organisms and ecosystems such as morphology, physiology, and cellular and molecular biology. Both perspectives converge, however, in their assumption that through systematic exploration of the biophysical elements of nature, living diversity can be comprehended and sometimes controlled.

Ecology is often viewed as a modern invention. Indeed, Aldo Leopold even called it "the outstanding scientific discovery of the twentieth century."[6] Nevertheless, an ecologistic perspective of species and habitats is undoubtedly quite ancient. Despite his assertion, Leopold intuitively recognized this possibility when he remarked: "Let no man jump to the conclusion that Babbitt must take his Ph.D. in ecology before he can 'see' his country. On the contrary, the Ph.D. may become as callous as an undertaker to the mysteries at which he officiates."[7]

Still, an ecologistic value tends to stress elements of nature and living diversity usually not evident to the average person. This lack of awareness may stem from the fact that most ecological processes depend on the functioning of obscure invertebrate and microbial organisms. Insects, for example, which constitute the majority of species, foster such critical ecological functions as pollination, seed dispersal, parasitism, predation, decomposition, energy and nutrient transfer, edible materials for other creatures, and the maintenance of biotic communities through various food webs. Most people are only dimly aware of these processes, let alone the creatures integral to their performance.[8] They tend to direct their attention instead to large vertebrates and other prominent features of the natural environment.

Despite this tendency, the intuitive recognition of ecological functioning has probably always been characteristic of the keen observer. Certainly an ecologistic outlook would confer distinctive advantages in meeting life's many physical and mental challenges—not only sharper knowledge and observational skills but the capacity to recognize material benefits from exploiting and mimicking natural processes. Understanding nature's functions and structures probably further instilled in the prudent person a cautious respect for maintaining natural systems and a reluctance to overexploit species and habitats. Perhaps this intuitive understanding explains why many aboriginal peoples often deliberately preserve much of pristine nature or why ancient India protected 6 percent of its land as sacred groves in contrast to a much smaller proportion protected today.[9]

The scientific perspective, by way of contrast, tends to place greater stress on the physical and mechanical functioning of living diversity. This viewpoint typically emphasizes the constituent elements of nature rather than entire organisms or relationships among species and natural habitats. Often the fundamental unit of analysis is the organ or cell. Despite this reductionistic tendency—often divorced from direct personal contact with the living environment—the scientific outlook shares with the ecologistic an intense respect for the systematic study of natural process. The ecologistic-scientific perspective may also reflect an intellectual satisfaction apart from any immediate practical utility. Yet the material advantages of this viewpoint are clear. One can imagine considerable gains conferred upon those who over the ages developed an especially refined capacity for precise observation, systematic analysis, and empirical study of even a fraction of life's varied complexity.

AESTHETIC

Nature and living diversity also exert an extraordinary aesthetic impact on people. Few characteristics of life so consistently arouse such strong emotions in people under so many circumstances. The complexity and power of the aesthetic response to nature are suggested by its wide-ranging expression from the contours of a mountain landscape to the ambient colors of a setting sun to the fleeting vitality of a breaching whale. Each aesthetic experience evokes a strong, primarily emotional, register in most people, provoking feelings of intense pleasure, even awe, at the physical splendor of the natural world.

Many people view the aesthetic response to nature as reflecting one's individual preference, as if each person and every culture cultivated its own unique sensibility. But the universal character of most aesthetic responses to living diversity suggests otherwise. Indeed, certain animals and landscapes appear to elicit consistent aesthetic responses in people under widely varying cultural and geographical circumstances. Few people dispute the beauty of a flowering rose, the majesty of a conical mountain, the grace of a flock of waterfowl in flight, any more than they differ about the aesthetic repugnance of a naked mole rat, a cold damp cave, or a fetid swamp. Moreover, certain aspects of nature appear to be key elements of the aesthetic response: vista, prospect, color, light, contrast, texture, movement, and others. These aesthetic elements seem to be associated with feelings of harmony, order, and an almost striving after an ideal.

The importance of the aesthetic value of nature is suggested by the inadequacy of artificial substitutes. Preference for natural design and natural pattern appears to be a deeply ingrained bias in the human animal. Ulrich, for example, after reviewing the research literature, reports: "One of the most clear-cut findings . . . is the consistent tendency to prefer natural scenes over built views, especially when the latter lack vegetation or water features. Several studies have [revealed that] even unspectacular or subpar natural views elicit higher aesthetic preference . . . than do all but a very small percentage of [manufactured] views."[10] Cross-cultural studies suggest similarly shared aesthetic preferences among people in different societies, although the research literature remains sketchy.[11]

Most people's aesthetic responses to nature tend to focus on large organisms, particularly mammals and birds. In contrast to the ecologistic-scientific emphasis on small and often obscure creatures, the aesthetic perspective stresses the "charismatic megavertebrates"—the bears, deer, wolves, antelopes, lions, cranes, swans, and so forth. Each value of living diversity appears to emphasize certain elements of the plant and animal world. The aesthetic outlook tends to place primary focus on the larger, more colorful, mobile, and diurnal species.[12]

Although the significance of the aesthetic emphasis on large creatures remains elusive, it may be a critical aspect of the tendency to impute value to certain landscapes. Perhaps the presence of an aesthetically salient creature animates, directs, organizes, and emotionally charges the human response. Conversely, the absence of certain species or landmarks may render the environment less focused, more haphazard, and even lifeless for the human observer. Leopold recognized the aesthetic significance of certain creatures in the landscape when he described how the ruffed grouse and wolf occupied for him this critical aesthetic niche (although for others this role might be filled by different creatures or natural features):

> The autumn landscape in the north woods is the land, plus a red maple, plus a ruffed grouse. In terms of conventional physics, the grouse represents only a millionth of either the mass or energy of an acre. Yet subtract the grouse and the whole thing is dead. An enormous amount of some kind of motive power has been lost. . . . My own conviction on this score dates from the day I saw a wolf die. . . . We reached the old wolf in time to watch a fierce green fire dying in her eyes. I realized then, and have known ever since, that there was something new to me in those eyes—something known only to her and to the mountain.[13]

The aesthetically salient animal or plant may constitute a central organizing element for the landscape—a focal point of meaning without which the terrain becomes undifferentiated, static, inanimate. This may help to explain George Schaller's reference to the great Himalayan mountains as "stones of silence," when he discovered the near extirpation of its most distinguishing fauna, the diversity of wild goat and sheep species.[14] Leopold, in contrast, invoked the organic simile of "thinking like a mountain" in arguing for protecting all the indigenous fauna of an ecosystem, even its most reviled creatures.[15] The aesthetically salient creature can confer on the landscape a critical vitality and order; its absence can transform the land into a mute and stony silence.

What is the advantage of the aesthetic experience of nature? It may reflect an intuitive recognition of an ideal modeled in nature: the magnificent stag, the mountain monarch, the brilliant butterfly, all suggest a striving after integrity, harmony, and balance in nature. Aesthetic attraction provides not only encouragement but templates of action for humans struggling to impose meaning and order on an existence filled with challenge and the potential for chaos. The aesthetic experience may point the way toward refinement and the possibility of unity and purposeful design in the animal and landscape in its idealized form. As Holmes Rolston suggests: "The aesthetician sees that ideal toward which a wild life is striving."[16]

The aesthetic attraction to varying species and landscapes may further reflect a recognition of the increased likelihood of finding food, safety, and security in nature. Gordon Orians, Judith Heerwagen, and others have reported that people consistently prefer landscapes with open vistas, water, and bright colors—all environmental features likely to offer sustenance and protection.[17] Singling out certain species and landscapes over the course of human evolution may have resulted in vistas, movements, resources, and other sensory cues being consistently preferred because they signified the increased probability of encountering safety and security.

The human aesthetic response to animals and nature suggests constancy and coherence rather than random fluctuation. More appears to be involved than a simple reaction to the pretty. Initially the aesthetic response to living diversity may seem casual, even trivial, but on closer inspection it embraces deeper levels of meaning. As Leopold tells us: "The physics of beauty is one area of natural sciences still in the dark ages. . . . Our ability to perceive quality in nature begins, as in art, with the pretty. It expands through successive stages of the beautiful to values as yet uncaptured by language."[18]

SYMBOLIC

The symbolic value reflects the human tendency to use nature for communication and thought. People have employed nature's rich tapestry of forms for expressing ideas and emotions for perhaps as long as humans have spoken. Through story, fantasy, and dream, the natural world offers raw material for building our species' seemingly unique and arguably most treasured of capacities: the ability to use language to exchange information among ourselves.[19] This use of nature represents the symbolic transforming of nature within ourselves, rather than our entering and engaging the natural world on its own terrain.[20]

Nature's symbolic value is most powerfully reflected in the development of human language. The complexity of language unfolds through the acquired ability to render refined distinctions and categorizations. The model for developing these complex differentiations and elaborate orderings largely resides in the natural world. Nature offers countless distinctions and opportunities for language development, especially among young children. Diversity in the natural world provides the growing child with a steady stream of objects for acquiring and practicing the tasks of ordering, sorting, and naming, all fundamental in the development of language skills. This function may be reflected in the finding that more than 90 percent of the characters in preschool children's language and counting books are animals and objects of nature. One giraffe, two bears, three elephants appears to represent a more compelling instructional set than one ball, two boxes, three chairs.

Animals seem to play a dominant role in the symbolic value of nature. This tendency may stem from our habit of imputing humanlike feelings and thoughts to these creatures. As Elizabeth Lawrence suggests: "The human need for metaphoric expression finds its greatest fulfillment through reference to the animal kingdom. No other realm offers such vivid expression of symbolic concepts. . . . It is remarkable to contemplate the paucity of other categories for conceptual frames of reference, so preeminent, widespread, and enduring is the habit of symbolizing in terms of animals."[21] Paul Shepard even more broadly asserts: "Human intelligence is bound to the presence of animals. They are the means by which cognition takes its first shape and they are the instruments for imagining abstract ideas and qualities. . . . They are the code images by which language retrieves ideas from memory at will. . . . They enable us to objectify qualities and traits. . . . Animals are used in the growth and development of the human person, in those most priceless qualities we lump together as 'mind.' . . . Animals . . . are basic to the development of speech and thought."[22]

Beyond language development, Shepard hints at other symbolic uses of nature. Story, myth, and fairy tale, often focused on animals and the natural world, have long assisted children in resolving dilemmas of selfhood, authority, power, and parental and societal relationship. Children's stories often use anthropomorphism—people disguised as animals—to help young people confront these poignant aspects of need, desire, and conflict in a compelling yet tolerable manner. This metaphorical device has been described by Bruno Bettelheim with regard to fairy tales, Joseph Campbell and Claude Lévi-Strauss in relation to myths, Carl Jung in the case of animal archetypes.[23] Each has revealed how natural symbols offer a means for confronting fundamental and often painful issues of identity, sexuality, and authority. Anthropomorphism, rather than being a distorted adulteration of the natural world, may reflect an indispensable device for human growth and development.

Another symbolic function of nature and living diversity employs terms from the natural world to facilitate everyday discourse. All languages use nature to render communication more vivid and persuasive, and many words have their origins in the natural world even when people are unaware of these etymological origins. This symbolic use of nature can produce poetry and vigorous discourse, as well as more mundane marketing and product promotion. What these various media have in common is their use of "thousands of figures of speech, phrases, colloquialisms, and neologisms" based on animal and natural imagery.[24]

Given modern society's enormous capacity for artificial fabrication, how much do humans today depend on the natural world for symbolic communication and thought? This question, of course, cuts across all the values of living diversity. Surely a few centuries of large-scale manufacturing could not have rendered irrelevant biologically encoded responses that evolved over hundreds of thousands of years when nature served as the only medium for human maturation. Moreover, the human emotional and intellectual need for varied distinctions even today appears to be matched only by the rich diversity and complexity encountered in the natural world. Artificial fabrications offer a meager substitute. Indeed, if manufactured artifacts represented people's only source of symbolic inspiration, the result might well be a stunted capacity for thought. As Elizabeth Lawrence ponders: "It is difficult to predict the ways in which our diminishing interactions with the natural world . . . will affect expressions of cognitive biophilia. . . . If we continue our current policy of destructiveness toward nature, does this mean that human language will contain fewer and fewer symbolic references to animals—with consequent impoverishment of thought and expression?"[25]

DOMINIONISTIC

Nature and wildlife have always confronted humans with significant challenges, physical and mental, testing and refining people's capacities for enduring, even mastering, the chore of survival in the face of worthy opposition. People have long contested the wild and, in the process, honed their ability to subdue and control the unruly and threatening elements of their world.[26] A dominionistic value can sometimes encourage an excessive urge to suppress nature, especially in our modern age of technology. Still, the relatively recent development of a frightening capacity to obliterate nature should not prevent us from recognizing the ancient and functional roots of this value.

Survival, even in the modern era, is still a tenuous enterprise necessitating some degree of human capacity for endurance and mastery. The ability to subdue, and the skills and prowess honed by an occasionally adversarial relationship with nature, remain essential ingredients in developing the human capability to survive. Perhaps this may explain why people often feel compelled to keep this aspect of the human spirit alive even when it seems superfluous. As Holmes Rolston suggests:

> The pioneer, pilgrim, explorer, and settler loved the frontier for the challenge and discipline. . . . One reason we lament the passing of wilderness is that we do not want entirely to tame this aboriginal element. . . . Half the beauty of life comes out of it. . . . The cougar's fang sharpens the deer's sight, the deer's fleet-footedness shapes a more supple lioness. . . . None of life's heroic quality is possible without this dialectical stress.[27]

The dominionistic experience of nature can sharpen mental and physical competence through testing various abilities and capacities. By successfully challenging nature and wildlife, people derive feelings of self-reliance that are hard to achieve in an untested relationship or by simply experiencing nature as a spectator.[28] The predator appreciates its prey to a degree no other creature can, and this may be as true for the human hunter of ducks and mushrooms as for the wolf stalking its deer or the deer seeking its browse. Although the dominionistic value may be less directly relevant today than in the human past, it would be a mistake to deny its legitimacy or reject the continuing desire to exercise mastery over nature. Like all the values, the dominionistic possesses both the potential for functional as well as exaggerated and self-defeating expression.

HUMANISTIC

Wildlife and nature also give people an avenue for expressing and developing the emotional capacities for attachment, bonding, intimacy, and companionship. For most people, these abilities are nurtured through close association with single species and individual animals, often culturally significant vertebrates and especially domesticated animals that become part of the human household. These creatures are frequently "humanized" in the sense of becoming companions and intimates not unlike other people.[29]

Although the humanistic value tends to be directed at animals with seemingly humanlike feelings and intelligence, this bond can sometimes be extended to creatures quite unlike people. Still, the humanistic stress on intimacy with individual animals usually precludes close relationships with species fundamentally different from people. Humanistic sentiments toward animals and nature can be so strongly manifest they correspond for many people with feelings of love. This degree of attachment usually focuses on domesticated pets. According to the historian Keith Thomas, what distinguishes the pet animal as a subject of strong affection is people's willingness to bring it into the home, give it a name, and forgo the possibility, even the thought, of eating it.[30]

So strong is the humanistic bond with companion animals they have been used in mental and physical therapies with disturbed, lonely, and estranged people.[31] This healing can also occur in the natural environment itself—reflected, perhaps, in the ancient tradition of seeking solace at seashores, mountains, deserts, and other prominent landscapes at times of acute distress. Conversely, pronounced feelings of loss, even bereavement, can be associated with the death of a companion animal, the destruction of certain landmarks, or the extinction of a species.

The humanistic experience of nature develops the capacities for caring, bonding, and kinship. As highly social animals, humans require these affective abilities, which increase the likelihood of cooperative, altruistic, and helping behavior so important to the survival of any social creature. Humans crave companionship. Emotional bonding with other creatures can satisfy this need and enhance our capacity to direct these emotions toward others. This sense of affiliation may also constitute an antidote to isolation and loneliness. Perhaps this is why the humanistic value has become so prevalent in modern society, as we shall see in the next chapter. Increasingly anonymous relationships, the erosion of extended family networks and stable communities, greater mobility—all have rendered close ties with other creatures more relevant than ever. As James Serpell suggests:

> Dogs and cats have maintained popularity as animal companions not, primarily, because they are home-loving, active during the day, nonaggressive or easy to house-train. . . . By seeking to be near us and soliciting our caresses, by their exuberant greetings and pain on separation, these animals persuade us that they love us and regard us highly. . . . People need to feel liked, respected, admired; they enjoy the sensation of being valued and needed by others. . . . Our confidence, our self-esteem, our ability to cope with the stresses of life, and, ultimately, our physical health depend on this sense of belonging.[32]

MORALISTIC

An almost incredible diversity characterizes life on earth. The number of species has been estimated at between 10 and 100 million or more, some 1.7 million species having been scientifically identified and described.[33] Known insects alone number more than 800,000 species with some 300,000 beetle and more than 100,000 ant and a similar number of butterfly and moth species. Moreover, the more we probe the mysteries of any one species or the structure of any particular ecosystem, the more we are astonished by the seemingly endless variety and complexity. Despite this extraordinary multiplicity of living forms and expressions, the essential similarity of all this diversity is equally remarkable. Much of life on earth seems to share a basic biomolecular structure and genetic character suggesting a common origin.[34]

This unifying structure has long suggested to many a basic symmetry, design, and even purpose underlying the natural world. For many this seeming unity has served as a source of spirituality, suggesting a fundamental order and harmony in nature, even a guide to human conduct. A moralistic value flows from discerning a basic kinship binding all life together, and an ethic emerges directing humans to minimize harm to other creatures viewed as fundamentally like ourselves—particularly species characterized by the seeming capacities for sentience, reasoning, and directed self-action. The moralistic value has been especially associated with concern for ethical treatment of animals and nature.[35] This value may involve strong affection for animals, but its more central focus is right and wrong conduct toward the nonhuman world.

A moralistic value has been associated with tribal peoples. Not only do such people frequently view the natural world as a living and vital being, but they also believe in an ethical reciprocity between humans, other creatures, and nature.[36] An analogous perspective is sometimes encountered among people in modern society—often rationalized by the link they see between solar energy, water, soil, plants, and animals. This modern articulation is reflected in John Steinbeck's words:

> It seems apparent that species are only commas in a sentence, that each species is at once the point and the base of a pyramid, that all life is related. . . . And then not only the meaning but the feeling about species grows misty. One merges into another, groups melt into ecological groups until the time when what we know as life meets and enters what we think of as non-life: barnacle and rock, rock and earth, earth and tree, tree and rain and air. And the units nestle into the whole and are inseparable from it. . . . And it is a strange thing that most of the feeling we call religious, most of the mystical outcrying which is one of the most prized and used and desired reactions of our species, is really the understanding and the attempt to say that man is related to the whole thing, related inextricably to all reality, known and unknowable. This is a simple thing to say, but a profound feeling of it made a Jesus, a St. Augustine, a Roger Bacon, a Charles Darwin, an Einstein. Each of them in his own tempo and with his own voice discovered and reaffirmed with astonishment the knowledge that all things are one thing and that one thing is all things—a

> plankton, a shimmering phosphorescence on the sea and the spinning planets and an expanding universe, all bound together by the elastic string of time.[37]

These moralistic sentiments of spiritual connectedness and ethical responsibility for nature have often been found in religion, philosophy, and the arts. The poet Walt Whitman, for example, remarked: "I believe a leaf of grass is no less than the journey-work of the stars . . . , and a mouse is miracle enough to stagger sextillions of infidels."[38] A moralistic perspective can increasingly be discerned, however, in the language of modern science.[39] Edward Wilson, for example, connects universal creation with modern biology when he suggests: "Biodiversity is the creation. . . . If humanity is to have a satisfying creation myth consistent with scientific knowledge . . . the narrative will draw to its conclusion in the origin of the diversity of life. Other species are our kin. . . . All higher eukaryotic organisms, from flowering plants to insects and humanity itself, are thought to have descended from a single ancestral population. . . . All this distant kinship is stamped by a common genetic code and elementary features of cell structure."[40]

A variety of advantages derive from a moralistic value of nature. This perspective may foster kinship, loyalty, and cooperation when powerfully articulated in a group context. Strongly held moralistic sentiments can also encourage the protection of living diversity viewed as replete with spiritual significance. But there is an adaptational advantage as well that derives from the sense of an underlying and unifying meaning, order, and purpose to life, particularly the confidence which flows from the conviction that a basic kinship binds all living creatures and the natural world together.

NEGATIVISTIC

So far the typology of values has emphasized positive and sympathetic relationships with animals and nature. Yet the natural world is also a powerful carrier of hostile and negative feelings: aversion, fear, and dislike, for example. Nature can evoke threatening and antagonistic sentiments to a degree as great as any encountered in the human experience. Certain animals and landscapes consistently provoke anxious reactions in many people under widely varying circumstances.[41] Snakes, spiders, sharks, scorpions, large predators, strong winds, stagnant swamps, dark caves—all can precipitate acute passions and avoidance responses in many people. And once aroused, these feelings are hard to extinguish.

These sentiments of dread and dislike can provoke destructive actions toward the natural world. But such fears can also encourage a healthy distancing and even respect for nature. Since avoidance of injury constitutes one of the most ancient motives of the human animal, a realistic avoidance of threatening aspects of nature is to be expected and at times welcomed. The advantage of isolating and even destroying potential danger in nature can be easily comprehended, and negativistic attitudes, within reason, may reflect functional behavior.

The tendency to avoid certain species is reflected in some people's reactions to large predators, snakes, and arthropods, particularly spiders and biting and stinging invertebrates. A predisposition seems apparent. Roger Ulrich notes: "Conditioning studies have shown nature settings containing snakes or spiders can elicit pronounced autonomic responses . . . even when presented subliminally."[42] Other researchers have observed that "ugly, slimy, and erratic" animals, such as snakes and spiders, frequently precipitate withdrawal responses among young primates (including human infants) even in the absence of an obvious threat.[43] These automatic reactions suggest a basic aspect of the human condition that might have conferred advantages during the long course of human evolution.

These fears can foster excessive, irrational, and extremely cruel behavior toward certain elements of the natural world. In some cases, negativistic sentiments can create an impulse to eradicate entire species. Barry Lopez implies that such loathing may have been an underlying urge in efforts to exterminate the wolf:

> Ever since man first began to wonder about wolves . . . he has made a regular business of killing them. At first glance the reasons are simple enough and justifiable. . . . But the wolf is fundamentally different because the history of killing wolves showed far less restraint and far more perversity. . . . Killing wolves has to do with fear based on superstitions. It has to do with duty. It has to do with proving manhood. . . . The most visible motive, and the one that best explains the excess of killing, is a type of

fear: theriophobia. Fear of the beast. Fear of the beast as an irrational, violent, insatiable creature.[44]

In the context of modern technology, negativistic sentiments can lead to the massive destruction of life. But this potential should not preclude our recognizing its biological origins or continuing advantage when manifest at a reasonable level. Fear of injury and death is still an integral component of human behavior. Moreover, dreading certain aspects of nature may be essential in developing a sense of awe, respect, and even reverence for the natural world. For nature has the power to humble, overwhelm, and even destroy human life. People would hardly manifest a healthy deference for the natural world if it lacked the capacity to frighten or intimidate. Has there ever been a god who did not possess the power to terrify as much as express love and compassion?

Biophilia

These nine values, considered biological in origin, signify basic structures of human relationship and adaptation to the natural world developed over the course of human evolution. Edward O. Wilson conceived the term "biophilia" to describe a deep biological need for affiliating with life and nature.[45] The nine values are thought to reflect a range of physical, emotional, and intellectual expressions of the biophilic tendency to associate with nature.

The values represent dispositions associated with the human inclination to affiliate with the natural world. These are weak biological tendencies, however, requiring learning and experience if they are to become stable and consistently manifest.[46] Unlike the "hard-wired" instincts of breathing or feeding, which occur almost automatically, the biophilic values must be cultivated to achieve their full expression. They depend on repeated exposure and social reinforcement before emerging as meaningful dimensions of human emotional and intellectual life. Once learned and supported, however, they become key elements of human personality and culture.

Learning and experience exert a fundamental shaping influence on the content, direction, and strength of these values. Without this reinforcement, the values may become vestigial and distorted. As Wilson remarks: "When humans remove themselves from the natural environment, the biophilic learning rules are not replaced by modern versions equally well adapted. . . . Instead, they persist from generation to generation, atrophied and fitfully manifested in the artificial new environments."[47]

In subsequent chapters we will consider the influence of learning and experience in shaping the values among diverse demographic and activity groups, in relation to different species, and in varying cultures and societies. Despite this variability, the underlying values remain constant. As complex biocultural phenomena, they indicate how people are influenced both genetically and culturally and how biological tendencies may become modified and channeled as a result of individual, group, and historical experience.

The concept of biophilia suggests these values of nature emerged because they conferred distinctive advantages to people in the process of evolutionary development. Human identity and fulfillment depend on the effective expression of these links to the natural world. Humans need to affiliate with nature and living diversity not just to ensure their material and physical well-being, but also to satisfy emotional, intellectual, and spiritual needs. The search for a coherent experience of self and society depends on the richness and variety of the human relationship to nature. Living diversity continues to serve as an essential medium for the developing person and is where the human species remains permanently rooted. The disintegration of these links diminishes our existence—not only materially but mentally as well. The human species can no more dissociate itself from the natural world than it can divorce itself from the products of cultural creation. Biological diversity and ecological process are the anvils on which human physical and mental fitness are formed.

An ethic of conserving nature and living diversity represents less an act of kindness than an expression of profound self-interest and biological imperative. Nature's healthy functioning reflects the human capacity to lead emotionally, intellectually, and spiritually rich and meaningful lives. As René Dubos remarks:

> Conservation is based on human value systems; its deepest significance is the human situation and the human heart. . . . The cult of wilderness is not a luxury; it is a necessity for the preservation of mental health. . . .

> Above and beyond the economic . . . reasons for conservation, there are aesthetic and moral ones which are even more compelling. . . . We are shaped by the earth. The characteristics of the environment in which we develop condition our biological and mental being and the quality of our life. Were it only for selfish reasons, therefore, we must maintain variety and harmony in nature.[48]

Hugh Iltis has similarly suggested: "Here, finally, is an argument for nature preservation free of purely utilitarian considerations; not just clean air because polluted air gives cancer; not just pure water because polluted water kills the fish we might like to catch; . . . but preservation of the natural ecosystem to give body and soul a chance to function in the way they were selected to function in their original phylogenetic home. . . . Could it be that the stimuli of non-human living diversity makes the difference between sanity and madness?"[49]

The claim of an inherent human need to value life and natural process could be criticized as a romantic ideology paraded in the guise of biology and promoted for essentially political purposes. This critique of the biophilia hypothesis might view this idea as condemning by implication all those mired in urban poverty or removed from contact with nature as victims of yet another stereotype of a less realized and fulfilling existence. The notion that people of lower socioeconomic status or city dwellers have less need to affiliate with nature ironically constitutes an elitist position. Nature's potential for fostering a satisfying existence may be less immediately apparent among the poor and urban than the rich and rural, but this represents more a challenge of design and opportunity than the irrelevance of the natural world for an entire class of people. Nature's capacity to enlarge the human experience is a power available to all but the most utterly deprived. Even the most impoverished city offers extraordinary opportunities for experiencing natural wonder. As Leopold has suggested: "The weeds in a city lot convey the same lesson as the redwoods. . . . Perception . . . cannot be purchased with either learned degrees or dollars; it grows at home as well as abroad, and he who has a little may use it to as good advantage as he who has much."[50] Society's challenge is to make the positive experience of nature accessible to all rather than to dismiss its presumed relevance to an entire group.

The more critical question is how much the human need to affiliate with nature has been debased and degraded by our massive destruction of life on earth. Modern society is destroying living diversity and natural process to an extent unrivaled in human history. Some estimates suggest that 15,000 to 30,000 species extinctions may be occurring annually due to the impact of human activities.[51] Possibly one-fifth of America's freshwater fish species and perhaps half of its mollusks have already become extinct or are in danger of extinction. Some one-fifth of the world's bird species have been eliminated, mainly because of human disturbance of island environments. Many of the world's large mammals, particularly its predators and carnivores, are endangered, as well as many marine mammals, especially the great whales. Known vertebrate extinctions in North America alone have increased from six in the 500-year period from 1100 to 1600, to twenty during the next 250 years from 1600 to 1850, to approximately five hundred during the past 150 years. The most traumatic extinction event is currently befalling the insects of the moist tropical forests. Wilson calculates that more than 20,000 species, mostly insects, may be eliminated annually, based on a tropical deforestation rate of somewhat less than 2 percent per year (assuming 10 million species, half in the tropics). The next fifty years could witness the disappearance of 100,000 species in North America alone. The possibility of a million and more extinctions worldwide represents a reasonable calculation during this time period. Moreover, species extinctions reflect only the tip of the iceberg, for many plant and animal populations will become severely depleted short of extinction. The drastic reduction of these species populations may seriously compromise many critical ecological functions and processes, as well as impair economic, emotional, and intellectual contributions these creatures make to human welfare.

The hemorrhaging of so much life on earth has principally resulted from widespread habitat destruction. The reasons for this loss are many: large-scale farming, mining, forestry, grazing, water impoundment and diversion, various forms of urbanization and industrialization, road and highway construction, and more. All these forces of human competitive exclusion of other life forms have been largely the consequence of increasing human numbers, technology, energy use, and per capita consumption of space and materials.

Apart from massive habitat destruction, another major cause of biodiversity loss has been the introduction, whether deliberate or accidental, of organisms and diseases to areas where they have never occurred before. The decline of many North American freshwater fish species, for example, can be traced to the introduction of nonnative species. A study of the Salt River in Arizona in 1900, for instance, identified fourteen native and no nonnative fish species; a sampling in 1920 revealed seven native species and two introduced fish species; by 1950, the native fish had disappeared altogether, replaced by some twenty invasive fish species.

Human overexploitation of species represents a third cause of the current extinction crisis. Despite increased recognition of the need to develop sustainable forms of wildlife utilization, the forces of human greed, uneducated affluence, modern transportation, population increase, and inept management continue to result in the overexploitation of many plant and animal species. Indeed, excessive exploitation today constitutes the primary factor in the endangerment of some one-third of all vertebrate species. Human uses of wild living resources for food, fuel, fashion, clothing, sport, medicine, and decoration continue to tempt people to consume substantially beyond the capacity of many species to replace themselves naturally.

Habitat destruction, nonnative species introductions, and overexploitation have sometimes converged to annihilate entire groups of biologically related creatures. A tragic example is the endemic birds of Hawaii.[52] Hawaii developed a particularly unique bird fauna because of its geographic isolation as the landmass most distant from any continent. At the time of Captain Cook's discovery of the islands in 1778, approximately eighty species of forest birds occurred in Hawaii and nowhere else. Some twenty-three are now extinct and at least thirty are endangered. The Hawaiian honeycreepers are particularly illustrative, their variation in bill shapes and body sizes especially indicative of the evolutionary processes described by Charles Darwin in the Galápagos Islands. Fifteen of the original forty-seven Hawaiian honeycreeper species are now extinct and another sixteen are endangered. The reasons for this loss of Hawaiian bird life include excessive hunting, habitat destruction, predation and disease caused by introduced species, fires, pesticides, feather gathering, and overcollection. Even conservation-minded hunters and birders introduced species—thinking they were contributing to Hawaii's animal life—because Hawaii, like other isolated island environments, contained many unique species but had relatively few in overall number.

Beyond the moral regret, does the prospect of 15,000 to 30,000 extinctions every year really constitute a significant problem? After all, this number of species still represents a small fraction of the earth's estimated 10 million species. Even if 15,000 to 30,000 extinctions occurred for the next thirty years, it would still constitute less than 10 percent of the total number of species. Moreover, the great majority of these extirpated species are likely to be obscure, even unknown, invertebrates of the distant tropics.

Fifteen thousand or even a million species may appear to be an expendable fraction of the earth's biological wealth, especially if their elimination serves to increase the standard of living of impoverished peoples and countries. Yet life exists not in isolation but as a matrix of subtle and profound connections, and from these relationships derive many critical ecological functions essential to the continuance of all life, our own included. These fundamental life support services include plant pollination and reproduction, waste decomposition, nutrient and energy cycling, various food webs, oxygen and water production, and more.[53] Our current knowledge only dimly recognizes the complexity of these ecological processes. Removing 10 percent or even 1 percent of the planet's species, or substantially reducing their genetic variability and interconnectedness, is a little like randomly destroying pieces of an extremely complex mechanism while blindly hoping not to damage some vital element or process. Reason suggests the cumulative effect of this destruction will inevitably compromise overall ecological functioning and structure. This is nothing less than a monstrous game of ecological roulette with potentially catastrophic impacts, particularly for future generations.

The current extinction crisis returns us to the original question: can we continue to experience lives rich in meaning and value if these lives are built on the destruction of so much of the earth's living heritage? This chapter suggests that the ultimate raw material for much of human industry, intellect, emotion, personality, and spirit originated—and remains rooted—in a healthy, abundant, and diverse biota. Human lives filled with meaning and purpose are unlikely to be sustained if they orbit about a depauperate world. People can survive the extirpation of many life forms, just as they may endure polluted water, fouled air, and contaminated soils. But will this impoverished condition permit people to prosper physically, emotionally, intellectually, and spiritually?

Diminishing the biological medium of our species' development will likely diminish the human condition. Living diversity remains integral to human functioning and well-being across a wide range of values. Beyond the ecological benefits, species also contribute food, medicine, clothing, industry, decoration, and other material products. All organisms reflect the adaptational consequence of millennia of evolutionary trial and error, representing unique physical and biochemical properties of potential significance to human comfort and survival. It would seem remarkably foolish to eradicate so much tangible gain just when human knowledge of genetics and bioengineering promises to expand our ability to benefit from much of this irreplaceable wealth. The earth's many and varied species constitute huge storehouses of untapped capital. Their gross eradication seems a little like burning the world's cash reserves to keep warm for a few winter days.

Aesthetic, humanistic, and symbolic values reflect the irreplaceable importance of living diversity, as well, but more for the development of human personality and culture. Our world would be immeasurably lonelier if wolves, bears, whales, tigers, elephants, pandas, eagles, giraffes, rhinoceroses, and so many other creatures no longer existed. Living diversity remains an essential element of human language, myth, and story, a vital source of our notions of beauty and understanding. The many creatures of the world inspire and instruct. They nurture us intellectually and enrich us emotionally. They provide us with a profound otherness for developing our knowledge of humanity, self, and society. To destroy these species is to replace a community of interest with a world of sadness, loss, and guilt. Kinship is replaced by isolation, beauty and grace by homogeneity and sameness, story and myth by enfeebled imagination and understanding.

Naturalistic and dominionistic values also represent critical wellsprings for human curiosity, creativity, competitiveness, and skill. The human experience of the trout in its stream, the eagle soaring on high, the grizzly bear in its wilderness—all feed our craving for discovery, adventure, and accomplishment. The challenges they represent would be hollow if we totally vanquished these creatures, or degraded them, rendering them so rare as to be practically unattainable or so tame as to be boring. Human ingenuity can never substitute for the challenge and vitality of experiencing a life in the wild. Virtual reality will always remain a vicarious fraud: a vain seeking to replicate the authentic.

Humans also derive sustenance from nature's capacity to render life spiritually meaningful. Kinship with living diversity confers a deep sense of connection to a larger and transcendent whole. These instructional ties would seem to be imperiled by so much destruction inflicted on the earth's living legacy. The moral shame that ensues can only corrode the human spirit. The ethics of extinction are difficult to comprehend in the abstract. The extirpation of a single creature can sometimes bring into greater relief the appalling finality of not just killing a life but destroying the very capacity of a singular species ever to be born again. Peter Matthiessen and Aldo Leopold capture this loss in reflecting on the extinction of two North American birds, the great auk and the passenger pigeon. Matthiessen remarks on the great auk's passing, the first animal to be rendered extinct at the hand of the European immigrant:

> For most of us, [the great auk's] passing is unimportant. The auk, from a practical point of view, was doubtless a dim-witted inhabitant of Godforsaken places, a primitive and freakish thing, ill-favored and ungainly. From a second and a more enlightened viewpoint, the great auk was the mightiest of its family, a highly evolved fisherman and swimmer, an ornament to the monotony of northern seas, and for centuries a crucial food source for the natives of the Atlantic coasts. More important, it was a living creature which died needlessly, the first species native to North America to become extinct by the hand of [European] man. . . . The concept of conservation is a far truer sign of civilization than that spoliation of a continent which we once confused with progress. . . . The finality of extinction is awesome, and not unrelated to the finality of eternity.[54]

Leopold eulogizes the last of the passenger pigeons, "Martha," the final representative of perhaps the most abundant bird species of modern times:

We grieve because no living man will see again the onrushing phalanx of victorious birds, sweeping a path for spring across the March skies, chasing the defeated winter from all the woods and prairies of Wisconsin. . . . There will always be [passenger] pigeons in books and in museums, but these are effigies and images, dead to all hardships and to all delights. Book-pigeons cannot dive out of a cloud to make the deer run for cover, or clap their wings in thunderous applause of mast-laden woods. Book-pigeons cannot breakfast on new-mown wheat in Minnesota, and dine on blueberries in Canada. They know no urge of seasons; they feel no kiss of sun, no lash of wind and weather. . . . Our grandfathers were less well-housed, well-fed, well-clothed than we are. The strivings by which they bettered their lot are also those which deprived us of pigeons. Perhaps we now grieve because we are not sure, in our hearts, that we have gained by the exchange. The gadgets of industry bring us more comforts than the pigeons did, but do they add as much to the glory of the spring?[55]

Part Two

Variations

CHAPTER 3

American Society

ALTHOUGH THE VALUES of living diversity have been depicted as biological tendencies, experience and culture, as noted, exert a profound influence on their content, direction, and intensity. The values may be easy to provoke and, once stimulated, they may be hard to extinguish, but they require repeated reinforcement to become stable and salient aspects of human personality and society. Language, custom, culture—all have a role in the formation of the values of living diversity. This chapter focuses on the appearance of these values in American society—first for the country as a whole and then among major demographic groups distinguished by education, income, age, gender, urban/rural residency, and ethnicity.

The previous chapter presented nine values of nature as a taxonomy for examining various views of nature and living diversity. Table 1 offers brief definitions of each of the values and summarizes their presumed benefits to the individual and society.

Table 1. A Typology of Basic Values

VALUE	DEFINITION	FUNCTION
Utilitarian	Practical and material exploitation of nature	Physical sustenance/security
Naturalistic	Direct experience and exploration of nature	Curiosity, discovery, recreation
Ecologistic-Scientific	Systematic study of structure, function, and relationship in nature	Knowledge, understanding, observational skills
Aesthetic	Physical appeal and beauty of nature	Inspiration, harmony, security
Symbolic	Use of nature for language and thought	Communication, mental development
Humanistic	Strong emotional attachment and "love" for aspects of nature	Bonding, sharing, cooperation, companionship
Moralistic	Spiritual reverence and ethical concern for nature	Order, meaning, kinship, altruism
Dominionistic	Mastery, physical control, dominance of nature	Mechanical skills, physical prowess, ability to subdue
Negativistic	Fear, aversion, alienation from nature	Security, protection, safety, awe

Methods

Much of the information presented here and in following chapters derives from studies conducted by myself and my colleagues over the past two decades. These investigations involved mainly surveys in which the values were measured by scales consisting of statistically clustered questions. Although the scales have been statistically corroborated, they represent only crude approximations of the underlying values. Surveys permit the efficient gathering of information from a large number of people, but they represent a blunt instrument for exploring the complexities of how people perceive nature. The limits of social surveys are illustrated by our inability to measure the aesthetic and symbolic values. Despite these problems, the study results provide a systematic basis for assessing the expression and distribution of most of the values of nature and living diversity.

Our studies relied on social-psychological techniques for estimating the different values. Translating these values into numerical terms involves specifying empirical criteria for each value and then developing a means for converting these criteria into quantifiable measurements. Social-psychological techniques use attitude questions focusing on a particular aspect of a value, offering a restricted number of response alternatives, converting these answers to numerical equivalents, and summing these scores across a series of statistically related questions to obtain an overall score.

The challenge of values assessment has provoked considerable debate, particularly when this information is used to formulate resource policy and assess environmental damages.[1] Biological assessments often employ more empirically valid indicators, but they are generally limited by their ability to measure only ecological and scientific values (though biological criteria can be used to help estimate aesthetic and naturalistic values). Biological assessments rely on such criteria as biological productivity, nutrient flux, biogeochemical cycling, species richness, genetic diversity, landscape heterogeneity, and biomass. The deficiencies of the biological approach include its lack of meaning to most people, its incommensurability with economic and social-psychological assessments, and its inability to measure most of the basic values of nature.

Economic estimations have often been used to assess environmental values. The popularity of economic measurement stems from its relevance to the marketplace and significance for most societies in the world today. As well, economic estimations represent an accepted way of comparing the different values. A major problem of economic assessment, however, is its bias toward values more amenable to monetary estimation—mainly utilitarian and other commodity-based benefits of nature. Ecological, scientific, aesthetic, humanistic, and moralistic values do not easily lend themselves to economic measurement, and thus they are often overlooked in economically oriented decision making—a tendency that has sparked contentious policy debates. Not only have attempts at developing economic measurements of noncommodity values been controversial, but they have often resulted in methodologies of questionable validity.

The problems of economic assessment of environmental values are often dismissed as trivial and likely to disappear when more effective techniques are developed. This optimism may be unfounded. Economic estimations are rooted in a materialistic perspective incapable of capturing the full range of physical, emotional, and intellectual functions and benefits of the various values of living diversity. Economic measurements can never fully apprehend, for example, the satisfaction derived from an atavistic immersion in wilderness—or the kinship sometimes achieved with other forms of life, or the ecological dynamics of a peat bog, or the astonishment at the prowess of a barnacle on a stormy seashore. These values do not translate well into the metrics of the dollar bill. Yet for many people these spiritual, aesthetic, ecological, and other experiences of nature are among the most-prized characteristics of the human condition. Although these values are exceedingly difficult to measure economically, they may very well possess greater significance than marketplace values. Social-psychological assessments, it is true, use limited and questionable measurement techniques. Moreover, the results are typically less accepted by decision makers. Still, this measurement strategy does a better job of capturing the widest range of values of nature and the living world. This approach, therefore, will be the basis for many of the findings discussed in this and subsequent chapters of the book.

A wide variety of surveys conducted over the past two decades furnish the data for estimating the values of living diversity. The most ambitious of these studies involved a nationwide investigation of American attitudes, knowledge, and behavior toward animals and nature. This 1980 investigation, conducted by me and my colleagues for the Fish and Wildlife Service, was based on one-hour personal interviews with more than three thousand Americans in the forty-eight contiguous states and Alaska.[2] This chapter draws from the results of this study, as well as more than twenty of my more recent investigations. Our results provide a broad glimpse of the distribution of these perspectives in American society. In short, they indicate how a complex country views the natural world at a particular moment in time. One must guard, however, against the temptation to view one group or individual as the reflection of an

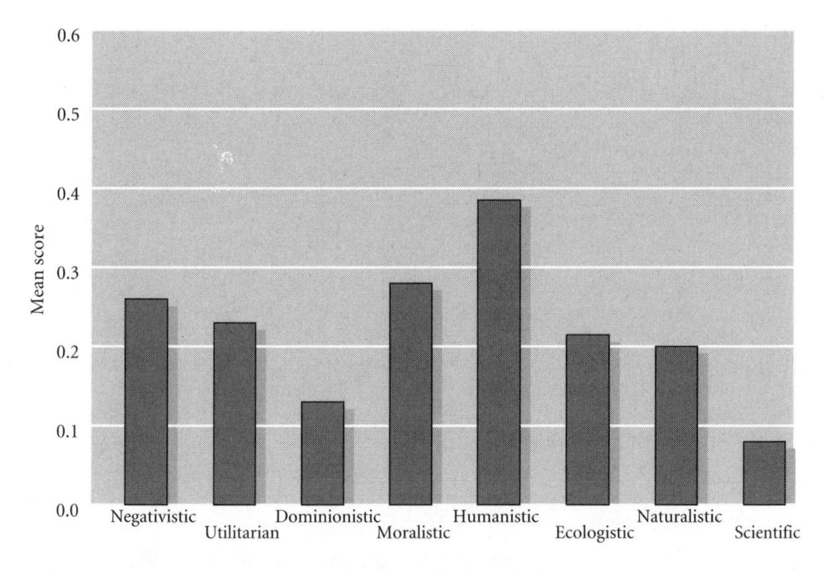

Figure 1. American Mean Attitude and Knowledge Scores

entire society or demographic group. Even when a large majority perceives nature in a certain way, a sizable minority may be inclined to think and behave quite differently.

Demographics

Figure 1 provides a rough estimate of the overall frequency of most of the values of living diversity in American society. These results indicate that the humanistic, utilitarian, moralistic, and negativistic values are the most prominently encountered perspectives. The most common view overall is strong affection for individual animals, especially pets; as for wildlife, there is a special preference for large charismatic species with strong cultural, historical, and aesthetic associations. A majority of Americans express pronounced feelings of attachment for pet animals. Most respondents also indicate a humanistic affinity for wildlife possessing physical and mental attributes frequently associated with humans—particularly animals of large size, considerable intelligence, familiarity, and the capacity for social bonding. The frequency of the humanistic value suggests a possible basis for the extraordinary number of pets in American society, the extensive popularity of zoos and national parks, and the prominence of nature and wildlife on television and film. Perhaps it also explains the widespread tendency of Americans to impute human motivation and characteristics to animals.

Despite this emotional attachment to certain animals and elements of nature, the negativistic value is also quite prevalent. Most responses to this value reveal less overt fear and dislike (except toward species like rats, bats, spiders, biting and stinging invertebrates, and others) and more indifference toward nature. Interestingly, negativistic sentiments are sometimes expressed by people with strong humanistic feelings for animals and nature: species falling outside their restricted sphere of affection are regarded as irrelevant and unappealing. These people might "love" pets, but they may harbor sentiments of apathy, even hostility, toward the broader realm of life.

A large proportion of Americans express strong moralistic concern for the proper treatment of animals and nature. Many object to various activities that presumably inflict suffering such as certain forms of trapping, trophy hunting, laboratory experimentation, and rodeos. Many also voice a willingness to forgo certain material benefits and recreational pleasures if it results in less harm and suffering to animals.

Despite these views, an impressive proportion of Americans expresses strong utilitarian values. Many place the material needs and interests of humans over nature and animals. These people emphasize the importance of exploiting the natural world and subordinating ethical considerations to more practical concerns. The importance of the utilitarian value is further suggested by a historical study covering the period 1900–1976.[3] According to this study, the utilitarian perspective remains the most prevalent value of animals and nature in American society, particularly among rural populations. Nevertheless, as Figure 2 suggests, with the exception of the two world wars, the utilitarian value has steadily declined in relative importance in American society.

The greater frequency of the humanistic, negativistic, moralistic, and utilitarian values helps to explain the often contentious debate in American society regarding the management and conservation of natural resources and wildlife. Emotional differences underlying these disputes may be reflected in a humanistic emphasis on strong attachments to in-

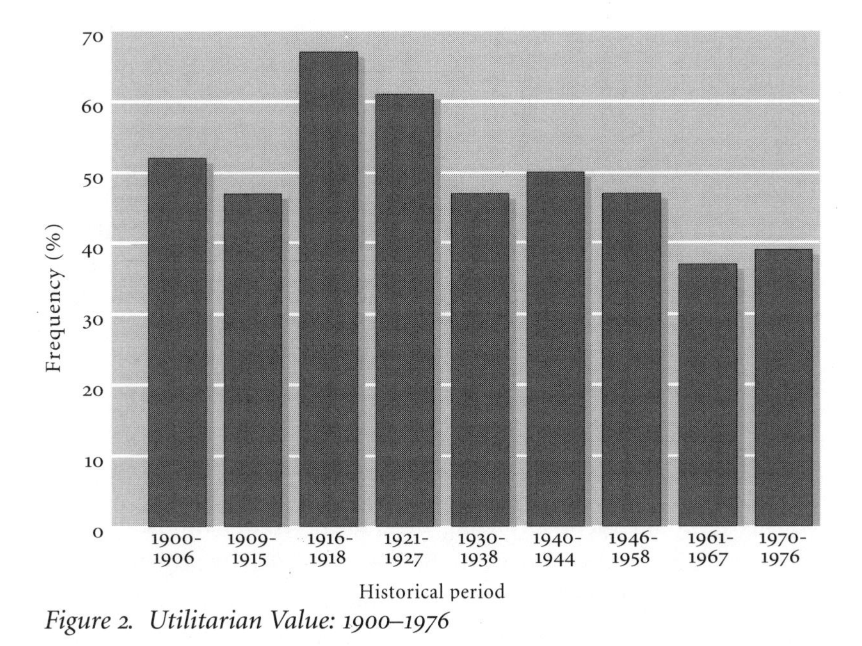

Figure 2. Utilitarian Value: 1900–1976

dividual animals versus a negativistic view of the natural world with indifference and sometimes animosity. The moralistic and utilitarian values further clash on matters involving the practical exploitation of animals and nature. The moralistic viewpoint often denies various human material benefits in the interest of protecting animals and the natural world; the utilitarian perspective typically regards such sacrifices as impractical and even foolish. The popularity of these four values of nature may represent an underlying basis for much contention regarding the appropriate management of wild living resources.

The least frequently encountered values of nature are the dominionistic, scientific, and ecologistic. The dominionistic value may have been more prevalent in American history, especially during the settlement period and through much of the nineteenth and even early twentieth centuries. The domination of nature during this time—particularly the conquest of its most formidable creatures, the bears, whales, wolves, cougars, and others—signified to many Americans the ability to take on a wild and untamed continent and, through intelligence and determination, master the land.[4] Today, however, the dominionistic perspective has become less relevant and, for some, a cause of regret. Perhaps this explains its declining popularity in contemporary American society, except among a small group of outdoor enthusiasts.

The scientific and ecologistic values are also expressed only rarely. The scientific emphasis on biological functioning, taxonomy, and physiological process appears to be of marginal interest to all but a few Americans. Although an ecologistic concern for natural history and organism–habitat relationships is more frequently encountered, it is still not common among the great majority of Americans. Satisfaction from the study of natural systems represents an unusual interest, but historical data (Figure 3) suggest it may be expanding. These findings, along with the humanistic and negativistic results, indicate that people have little interest in unfamiliar animals and unusual aspects of the natural world. Perhaps it should not be surprising, then, that a recent national study we conducted reveals most Americans had never heard of the concept of biological diversity or its loss and, as the results of Chapter 5 will demonstrate, remain largely unaware of most invertebrate life.

Naturalistic and aesthetic values of nature and living diversity are expressed only moderately. The desire for direct contact with animals and the natural world, particularly in a recreational context, is especially prominent among younger and better-educated people but still less manifest than the humanistic, negativistic, utilitarian, or moralistic values. Naturalistic perspectives seem quite variable among the American population: a minority are intensely interested in direct enjoyment and experience of wildlife and nature, while many view this prospect with indifference. Although we were unable to devise an adequate aesthetic scale, other data suggest that satisfaction derived from the physical appeal of animals and nature is widespread among Americans, though strongly manifest only among a few.

Given the limited expression of ecologistic and scientific values, it may not be surprising how little Americans know about living diversity. People's factual knowledge tends to be most apparent when it involves familiar creatures or animals commonly associated with human injury or property damage. Moreover, most people obtain their understanding of animals from everyday experiences rather than from books or formal education. Most recognize the names of widely known poisonous snakes,

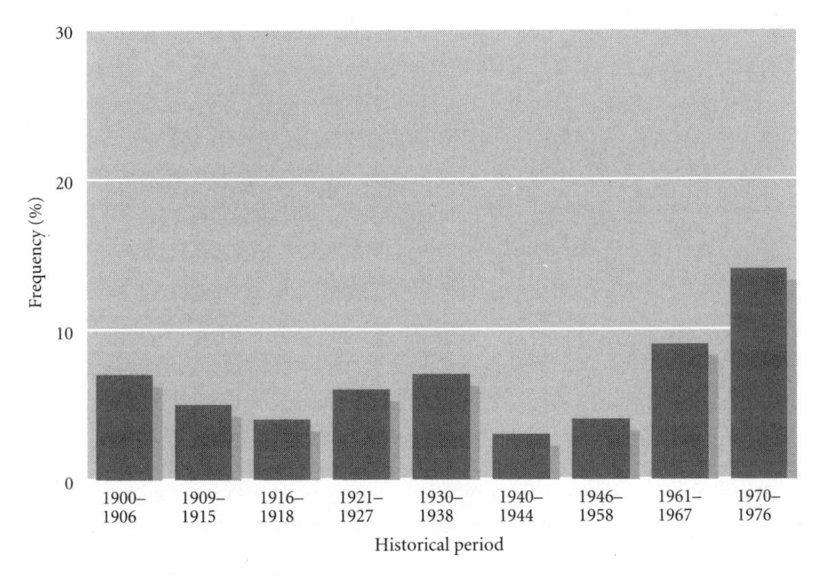

Figure 3. Ecologistic Value: 1900–1976

for example, but few understand the basic biological characteristics of snakes or their typical way of moving. (Most people in the survey believed snakes are covered with a thin layer of slime in order to move more easily.) Only half of the American population recognizes that invertebrates lack a backbone, and most think insects maintain a constant body temperature like birds or mammals. People reveal more knowledge of domestic animals and less knowledge of wildlife. Biological understanding is especially deficient when it involves insects, endangered species, and animal taxonomy and classification. (Only a slim majority recognizes beetles as the most common type of animal, turtles as unrelated to snails, or manatees as not being an insect.)

Although most people derive their knowledge of the natural world from personal experience, teenagers are an exception. Young persons reveal substantially greater knowledge than adults of animals associated with formal learning, while adults display greater understanding of nature based on practical experience. Teenagers routinely demonstrate superior knowledge of animal taxonomy or unfamiliar taxa such as insects, for example, whereas adults reveal greater knowledge of foods obtained from animals, common pests in the garden, and environmental threats associated with agriculture. Young Americans appear increasingly to get their knowledge of the natural world from books and schools; adults tend to rely more on personal experience.

This variability among adults and teenagers illustrates many differences found among groups of Americans in their relative values and knowledge of animals and the natural world. Because America remains a heterogeneous society, generalizations about the population as a whole are often problematic and misleading. Moreover, people rarely reveal just one or a small set of values; instead, they tend to express a hierarchy of perspectives ranked from primary to tertiary significance. The remainder of this chapter explores these differences in values and knowledge of nature and wildlife among Americans distinguished by age, gender, education, income, urban/rural residence, and ethnicity.

Age

Do values of nature vary among age groups? This is a question we explored through separate studies of children and adults. The children's investigation involved four age groups and focused on the developmental emergence of the values; the adult study emphasized value orientations among Americans as they moved through the life cycle. The influence of age and other demographic factors is also considered in subsequent chapters when we discuss particular wildlife issues, species, and cultures.

Jean Piaget, Lawrence Kohlberg, and others have noted distinct developmental stages in the emergence of children's intellectual and emotional thinking.[5] These studies have focused on children's perspectives of other people, rather than nature, and the adaptational requirements of living in a world dominated by human culture. As these studies are extraordinarily detailed and complex, we can consider them only briefly here. Essentially this research found clear developmental stages in children's thinking indicative of an evolving capacity to assimilate and accommodate new experiences. The typical child progresses from highly concrete to more abstract and logical reasoning and from largely self-centered thinking to more socially oriented feelings toward other people and human groups. Piaget noted three distinct stages in the development of logical reasoning capacities in the young child; Kohlberg identified six parallel stages in children's moral development.

Independent of these findings, my colleague Miriam Westervelt and I encountered an analogous developmental sequence in the emergence of children's values of nature and animals. An important difference, though, seems to be the later development of nature-related values in children—perhaps indicative of the lesser significance of human/nature versus human/social relations in the normal developing child, at least in modern society. We noted four distinct stages in children's developing values of nature and living diversity, reflected in the results of Figures 4 and 5.[6] Consistent with Piaget and Kohlberg's findings, children under six years of age were found to be egocentric, domineering, and self-serving in their values of animals and nature, a tendency reflected in especially high utilitarian and dominionistic scores. Young children reveal little recognition or appreciation of the autonomous feelings and independence of animals. They also express the greatest fear of the natural world and indifference toward all but a few familiar creatures. These young children further demonstrate, not unexpectedly, the least factual knowledge of animals and nature.

Initially we were surprised by the results. Our society frequently romanticizes young children's attitudes toward animals, believing that they possess some special intuitive affinity for the natural world and that animals constitute for young people little friends or kindred spirits.[7] As Piaget and Kohlberg have demonstrated in another context, the world of young children is often dominated by great fears and great needs, and self-centered thinking and concrete gratifications often take precedence over other considerations. Our results similarly show that young children typically view animals and nature in highly instrumental, egocentric, and exploitative ways, largely responding to short-term needs and anxiety toward the unknown.

The most dramatic value shift among children between six and nine is a decided increase in appreciation of the autonomy and independence of other creatures. During this second stage, children become more aware of animals as possessing interests and feelings unrelated to themselves. These young children also begin to recognize wild animals—especially

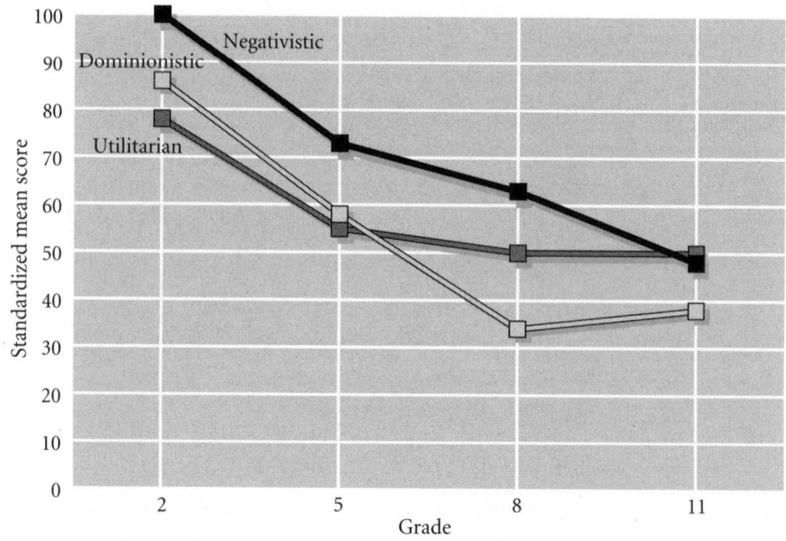

Figure 4. Attitudes That Decrease with Age: Children's Negativistic, Humanistic, and Dominionistic Values

Figure 5. Attitudes That Increase with Age: Children's Ecologistic, Moralistic, and Naturalistic Values

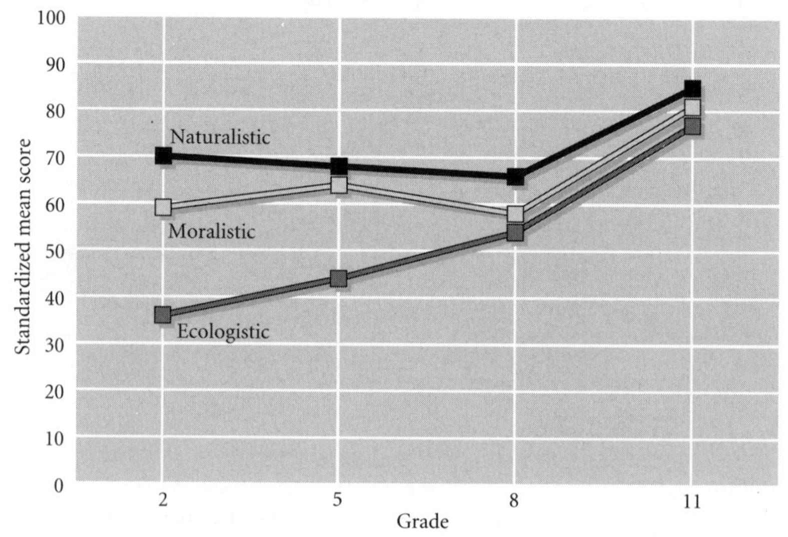

creatures seemingly like humans in behavior and intelligence—as having separate interests and desires. These children further begin to realize animals might suffer pain and distress. They consequently develop a conscience toward the nonhuman world, recognizing animals and nature as having the right not to be selfishly manipulated, a view motivated by more than just the possibility of being punished for harming other creatures.

The third stage in children's developing values of nature occurs between nine and twelve years of age. The most pronounced change involves a dramatic increase in factual understanding and knowledge of animals and the natural world. Unlike the feeling-oriented focus of the first two stages, this third stage emphasizes cognitive perspectives of nature and living diversity. This phase of early adolescence reflects rapid acquisition of knowledge, skills, and understanding of animals and the natural world. This period marks a bolder interaction of children with nature—a time for testing and exercising curiosity, interest, and fascination for the nonhuman world and how it functions.

The fourth and final stage in children's developing values of animals and nature occurs between thirteen and seventeen years of age. This period witnesses a sharp increase in abstract, conceptual, and ethical reasoning about the natural world. It is at this stage that moralistic and ecologistic values become prominent. These older children begin to comprehend relationships among creatures and habitats, as well as people's ethical responsibilities for exercising stewardship toward the natural world. Many teenagers become acutely concerned with conservation and treating other creatures with moral consideration. These older adolescents also reveal substantially greater naturalistic interest. They seek direct experience with wildlife and exploration of challenging natural environments.

These results suggest the importance of age in educating children about animals and the natural world. It seems pointless to focus on teaching very young children ecology and ethical responsibilities for conserving nature at a time when they are incapable of internalizing this type of abstract and compassionate thinking. Instead they might be encouraged to develop an emotional appreciation for animals and recognizing nonhuman creatures as possessing feelings and needs of their own. Middle and early adolescence, however, seems to be the most favorable time for emphasizing rapid acquisition of factual knowledge and understanding of the natural world. Finally, late adolescence is the best time for learning about the complexities of ecology and ethical responsibilities for the natural world.

Some similarities and differences among children and adults should be mentioned here before we consider adult values toward nature and animals. Both children and adults reveal strong humanistic perspectives of the natural world—particularly pronounced affection for individual animals, higher vertebrates, and domesticated pets. Children and adults similarly express strong negativistic values, especially fear of invertebrates, snakes, and certain wilderness settings. But there are substantial differences too. Children generally express less concern than adults for the practical benefits of nature, whereas adults show less interest than young people in direct experience of the outdoors and wildlife.

In our study of adult values we focused on persons eighteen years of age and older. The most outstanding difference occurred among persons eighteen to twenty five years of age and elderly Americans. Older people generally express significantly less interest and affection—and more dominionistic and utilitarian values—toward animals and nature than do young adults. The elderly also reveal consistently greater support for placing human economic and social interests over wildlife protection or the recreational enjoyment of nature. In contrast, young adults express strong moralistic values, often supporting the rights of nature over the economic interests of people, and tend to reject the assumption of human mastery of the natural world. Despite these considerable variations in perspectives, we found no significant differences among young adults and the elderly in their knowledge of animals and nature.

When viewed along with the children's findings, these results suggest a curious symmetry among the very young and old in basic values of the natural world. These two age groups reveal the greatest tendency to subordinate the interests of nature for the benefit of people and express the least support for moralistic and ecologistic values. Late adolescents and young adults, in contrast, show the strongest interest in conserving nature, particularly strong moralistic support for protecting animals and the natural world. Perhaps the dependent status of early childhood and old age fosters an enhanced appreciation of the utility of nature, as well as fear of threatening aspects of the natural world. Late adolescence and early adulthood, by way of contrast, might reflect a period of heightened curiosity for the unfamiliar, a belief in one's capacity to confront the challenges of the external world, and a time of surging optimism about what can and should be done in managing nature.

We were unable to tell whether these differences among young and older adults reflect basic changes in American society or normal shifts in the life cycle. In other words: do variations in values among the young and elderly indicate the cultural development of more appreciative attitudes toward the natural world, or are they simply a reflection of increasing conservatism as people age and confront the burdens of family, work, and security? Although definitive answers cannot be offered, other historical data suggest that both tendencies may be occurring. Differences among young and elderly Americans appear to suggest substantial changes in American society as this country becomes more conservation oriented and begins to favor protectionist values toward animals and the natural world. It also appears that the onset of family and career responsibilities engenders more pragmatic, exploitative, and less idealistic perceptions of nature and living diversity.

Gender

When we assessed male and female values and knowledge of nature, major differences emerged. These variations are reflected in Figure 6. Women consistently express greater humanistic and moralistic sentiments toward nature—particularly strong affection and emotional attachment to individual animals, especially pets. Women also reveal greater moralistic concerns than men, illustrated by less support for practices presumably involving substantial harm and suffering inflicted on animals including hunting, trapping, and various wildlife "harvesting" activities. Women are also much more likely to join groups opposed to the consumptive use of animals, whereas men account for most of the members of hunting and fishing organizations. Men express far greater support than women for the practical exploitation and domination of animals and nature. Men also reveal a greater intellectual

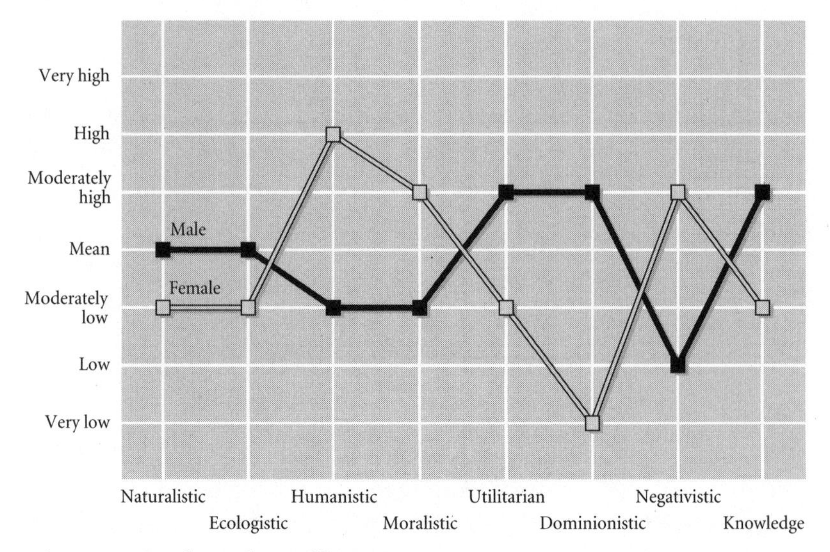

Figure 6. Gender Value Differences

orientation to the natural world, suggested by substantially higher scientific, ecologistic, and knowledge scores. Moreover, men express greater interest in outdoor recreation. Finally, men reveal significantly more interest and less fear of wildlife and nature, despite the tendency of women to express more emotional attachments to individual, especially domesticated, animals.

The size and strength of these gender differences are striking. In trying to understand their basis, my colleague Joyce Berry and I encountered the work of the developmental psychologist Carol Gilligan.[8] Gilligan had explored moral and cognitive development among women after having found shortcomings in the findings of Lawrence Kohlberg and Jean Piaget because of their reliance on data derived from studying mainly men.[9] Gilligan reports a greater moral tendency among females than males to stress the role of caretaker and a related inclination to obtain satisfaction from personal relationships involving a matrix of others, particularly family members, close friends, and local communities. Men tend to rely less on personal relations, preferring instead to emphasize work and competition in defining and promoting their sense of worth. Essentially Gilligan finds men far more inclined than women to structure the world according to logical, rational, and abstract rules of conduct.

Men stress the importance of positional and hierarchical ties, the distribution of power and authority, and maintaining "rights" based on formally articulated rules of conduct. Gilligan reports that women prefer to achieve personal meaning through caring and intimate relations with others, especially family and friends. Unlike the male stress on relatively fixed ideas of fairness, logic, and rights, females tend to emphasize interpersonal responsibilities, compassionate relationships, and the importance of affection and familiarity.

Gilligan's findings parallel the variations we encountered in our research regarding men and women's values of nature and living diversity. Consistent with Gilligan's finding of a female emphasis on caring and affection, we encountered a humanistic stress among most women on strong emotional attachments to individual aspects of nature along with a moralistic inclination to protect animals from harm. The unusual combination of humanistic and negativistic values among women also seems consistent with Gilligan's finding of strong ties with family and friends but relative disinterest in remote relationships. Men's greater stress on logic and hierarchical values of nature also appears consistent with our findings of the importance among men of mastery and control, practical utilization, and abstract and intellectual understandings of the natural world.

These results may help explain widespread differences among men and women in attitudes toward various wildlife and natural resource issues. Gender variations often emerge in disputes about the harvesting of furbearers, about predator and animal damage control (denning and poison control techniques), about harvesting practices (such as leg-hold traps), and other issues. While legitimate differences do separate those for and against these activities, basic value distinctions often underlie these conflicts. And these variations may be related to demographic differences such as those between men and women. Traditionally, the wildlife management profession has been overwhelmingly male, and this may explain the wide divergence in views separating wildlife managers from various animal welfare organizations, whose members are largely women.[10] Wildlife conservation must derive from the best science available, but effective policy should also consider the values of various stakeholders. The results presented here suggest that differences among males and females may be relevant considerations and should be acknowledged in developing equitable policies for managing nature and biological diversity.

Education and Income

Higher socioeconomic status usually confers greater leisure time and mobility, more opportunity for enjoying the recreational advantages of the natural world, and less financial dependence on deriving a living from exploiting natural resources.[11] It may not be surprising, then, that substantially higher naturalistic and lower utilitarian scores appear among people of higher income and education. Still, the relationship of socioeconomic status with values of nature is rarely simple—pronounced differences, for example, emerge among people of varying education and income levels. Education generally is far more strongly associated than income with emotional and intellectual perspectives of animals and the natural world. Education, in fact, emerges as the most powerful force shaping perceptions of nature and living diversity.

Differences among adults of varying education are reflected in Figure 7: nearly a mirror image of contrasting views between the least and most educated. The higher a person's education, the more likely that person is to express greater concern, affection, interest, and knowledge (and less exploitative and authoritarian attitudes) toward animals and the natural world. This tendency is especially pronounced among the college-educated. Moreover, even after distinguishing the college-educated by their disciplinary concentration in college, no significant differences emerge among humanities, social science, or science majors. College education, whatever the concentration, appears to foster sentiments of appreciation, concern, and understanding of nature and living diversity. Perhaps this tendency reflects the impact of "deferred adolescence" and delayed entrance into the work force, enabling people to internalize a more benign and less exploitative relationship to the natural world. College education may also promote greater knowledge and sense of stewardship toward nature and animals. Whatever the explanation, these results reflect the progressive impact of higher education. People with less than a high school education reveal significantly less understanding, more exploitative attitudes, and generally less sympathy

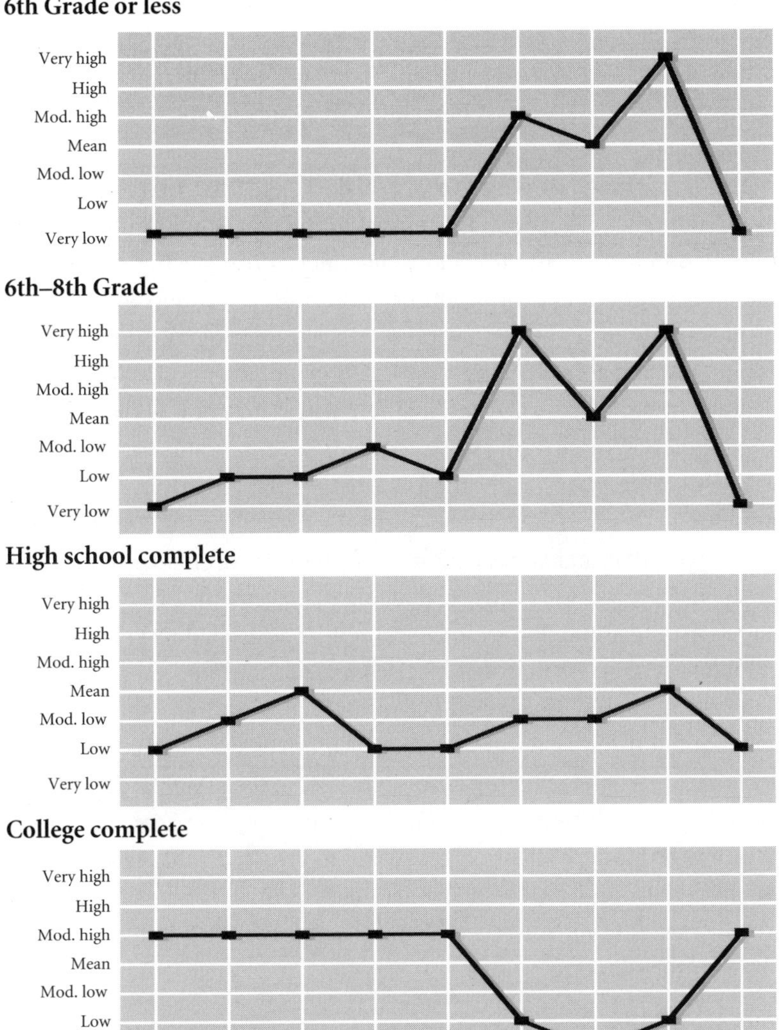

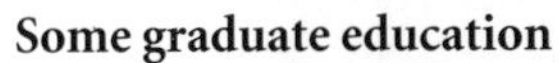

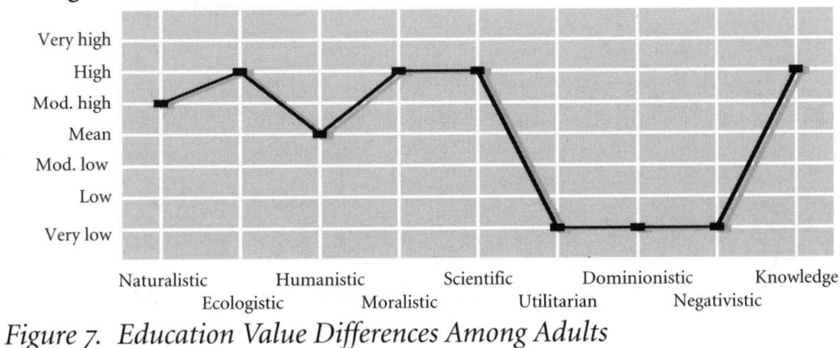

Figure 7. Education Value Differences Among Adults

toward nature and its conservation. Even after controlling for the possible confounding effects of other demographic variables, these results remain.

The importance of education is further suggested by the less pronounced impact of income. In contrast to the education results, no significant differences emerge among income groups on the ecologistic, utilitarian, moralistic, or dominionistic scales. People with higher income are no more likely to express ethical or ecologistic concerns (or less interest in exploiting and dominating wildlife) than people of low income—despite their substantial differences in outdoor recreational interest and knowledge of the natural world.

These results have implications for educating Americans about the importance of a healthy, diverse, and abundant biota. At the very least they suggest that higher education can foster a more enlightened regard for the natural world and its sustainable management. They also imply that environmental education should not be restricted to grade school and high school. The absence of appreciable differences among the college-educated distinguished by major further suggests the potential benefit of infusing environmental education across the college curriculum rather than isolating it, for example, in the sciences. Higher education seems to have a profound influence on people's perception of the natural world independent of its relation to greater wealth, leisure time, and material opportunities. The message seems clear: society should provide meaningful contacts with nature and wildlife for all Americans whatever their income.

Urban/Rural

Urban/rural variations appear to remain a fundamental factor in American perceptions of the land and its creatures.[12] The traditional distinction between those residing in small towns and those living in big cities has, however, altered substantially. The suburbanization of the American countryside as a consequence of major developments in transportation and communications has resulted in a striking transformation: once insular, rural, and agricultural communities have seen an influx of people who maintain an urban lifestyle and profession. The population of a person's hometown is, thus, no longer a reliable indicator of traditional urban/rural distinctions. The extent of dependence on the land and its resources for a livelihood appears to constitute a better predictor of environmental and wildlife values than the size of one's township.

Reflecting this change in traditional urban/rural distinctions, few differences in values of nature emerge among Americans distinguished by population of residence. But there were large and consistent variations depending on the amount of land people owned, whether they resided in large cities versus open country areas, or whether they were employed in resource-related professions such as farming, logging, commercial fishing, and mining. These urban/rural differences were significantly related to nearly all the values of nature and living diversity. Utilitarian and moralistic differences are a good case in point (Figures 8 and 9). Farmers, people who own large amounts of land, and those residing in open country areas express far more pragmatic (and less protectionist) attitudes toward nature and animals than do residents of large cities, people who own little or no land, the college-educated (likely to be engaged in urban white-collar professions), and young adults. Rural-oriented groups tend to endorse people's right to exploit and master nature; urban and suburban young and better-educated people express greater concern for the protection of wildlife and natural habitats. Urban/rural differences are especially pronounced when we examine where people were raised rather than where they currently reside. Striking utilitarian and moralistic differences emerge among persons raised in open country as opposed to large cities of more than a million people.

Despite these pronounced urban/rural differences, we find no significant variations in knowledge of nature and animals. Fundamental variations in basic perspectives of nature often have no connection to factual understanding. This finding should caution one against assuming that simply providing people with more knowledge will alter their attitudes toward nature and living diversity. More often than not, people employ new knowledge to reinforce their existing biases rather than to change strongly held views.

Urban/rural differences are often encountered in debates over the conservation, protection, and management of biological diversity. Later chapters will note these urban/rural variations in relation to such issues

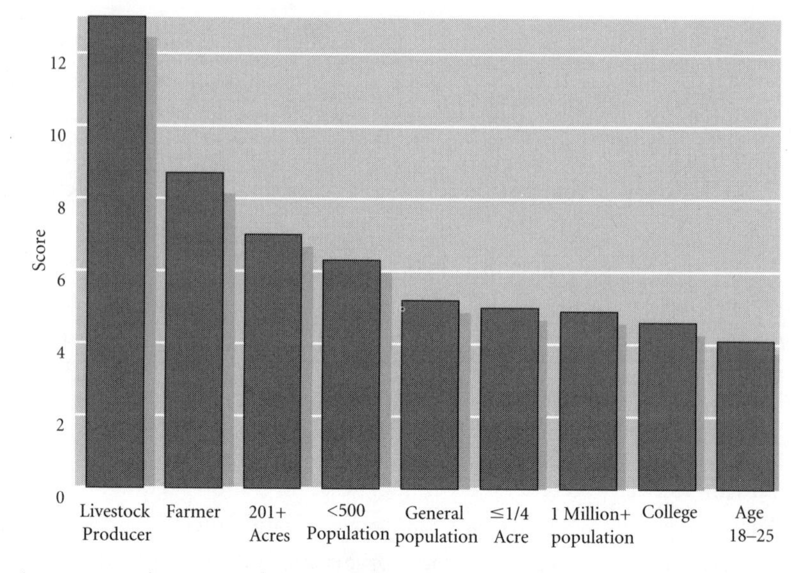

Figure 8. Utilitarian Urban/Rural Value Differences

Figure 9. Moralistic Urban/Rural Value Differences

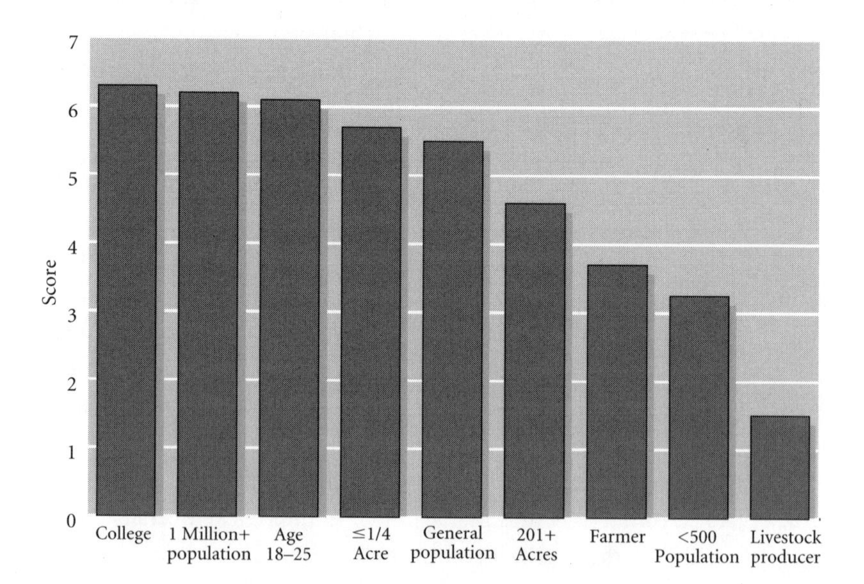

as exploiting natural resources, controlling injurious wildlife, making trade-offs between human development and species conservation, and protecting and restoring endangered species. As we shall see, resource-dependent populations consistently favor the utilization, subordination, and control of nature, while urban and suburban populations express greater support for the nonconsumptive use and protection of wildlife. Although legitimate differences often distinguish these urban/rural positions, behind them one often finds contrasting values of life and nature, particularly those developed during childhood. Extracting a living from the land often encourages a belief in the need to subordinate nature, whatever affection for the natural world one might have. Many rural people have a deep affinity for the land and its creatures, but they tend to view these resources from the perspective of their utility and a familiarity that often takes their long-term welfare for granted.

Many city residents, by way of contrast, wear a different set of blinders toward nature and wildlife. A highly romantic appreciation of the natural world frequently prevails among urban people, leading them to view as irrelevant and sometimes contemptible the practical dependencies on mastering wild living resources. In their eyes, nature may be viewed simplistically as large, attractive, and charismatic species or perhaps as pristine wilderness whose innocence and grandeur are debased by the hands of humans. For many urban wildlife advocates, practical considerations of animal harvest, control, and utilization are often viewed as unappealing and even repugnant. Nature as resource or challenge is inclined to be regarded with indifference, secondary to more aesthetic and nonconsumptive recreational interests. Ironically, many city residents barely recognize nature as an integral part of their urban lives—either from the perspective of practical benefits derived from environmental exploitation or in the natural complexity encountered even in the largest cities.

American society appears increasingly divided by the contrasting environmental values of urban and rural residents. Effective conservation of nature will necessitate a balanced consideration of both viewpoints. Differences among rural and urban dwellers often reflect legitimate perspectives, but sometimes they also reflect bias and intolerance. Wildlife management must strike a balance between the interests of all stakeholders, regardless of residence, while minimizing their distorted views toward the natural world.

Ethnicity

Significant differences also emerge among varying ethnic groups, especially those of European and African descent. Data on ethnicity are often lacking and inconsistent, however, and these results should be only tentatively accepted. Various studies, nonetheless, suggest that African-Americans evince substantially less interest, concern, and knowledge than European-Americans for nature and living diversity.[13] This impression is reflected in Figure 10, particularly the ecologistic, naturalistic, moralistic, negativistic, and knowledge results. African-Americans generally reveal less appreciation, less recreational interest, and less willingness to support the protection of nature and wildlife. They also express a greater inclination to endorse the practical exploitation and control of the natural world. Yet no significant differences emerge among European-Americans and African-Americans in their affection for individual animals or scientific interest in the natural world.

Any assessment of ethnic differences in American society must consider the distorting influence of unequal income and education. When this confounding influence is examined, however, we find the opposite of what might be expected. Higher socioeconomic status, rather than narrowing differences among European-Americans and African-Americans, actually increases them. The differences between African-Americans and European-Americans with less than a high school education are insignificant on the naturalistic, negativistic, and knowledge scales. Among college-educated blacks and whites, however, there are sharp value differences on these and other scales. College-educated African-Americans reveal limited appreciation, interest, concern, and understanding of living diversity and its conservation. Few significant value differences emerge among urban and rural African-Americans, again in considerable contrast to the pattern found among European-Americans.[14]

These findings suggest that most African-Americans view the experience and conservation of nature as being of only marginal concern. An explanation of this finding may be provided by the 1960s African-

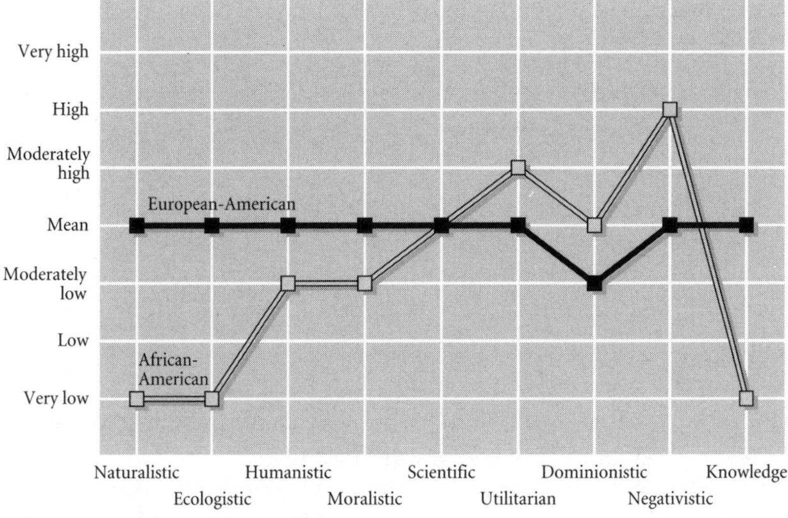

Figure 10. Ethnic Value Differences

American leader, Eldredge Cleaver, who suggested the experience of slavery was so traumatizing that African-Americans ever since have tended to view their progress in terms of distance from the land and its resources, frequently seen as symbols of exploitation, dominance, and exclusion.[15] Cleaver remarked: "Blacks learned to hate the land and came to measure their own value according to the number of degrees . . . away from the soil." African-Americans of higher socioeconomic status may gauge their success in terms of more urban than rural symbols, viewing the urban setting as the best site for upward mobility and recreation. The historical persecution of blacks and other minorities in association with rural areas may also elucidate why relatively few nonwhites visit national parks and other protected natural areas, perhaps largely viewed as pleasure grounds for European-Americans.

It appears that a large proportion of African-Americans do not place a particularly high value on the positive experience of living diversity, nor do they strongly support its protection. Wildlife remains for many African-Americans a peripheral issue. Indeed, ethnic differences may constitute an Achilles heel of the conservation movement, especially hindering its ability to make a convincing case for protecting diminishing natural resources and rediscovering human emotional and intellectual dependence on nature and living diversity. Until all ethnic groups believe that the chances for leading a richer and more rewarding life depend on a healthy, diverse, and abundant biota, this country may not be able to elicit the commitment necessary to halt the current mass destruction of life on earth.

The findings presented here are admittedly based on limited evidence. Moreover, recent research suggests that younger African-Americans confronted with enjoyable outdoor experiences often become substantially more appreciative, interested, and concerned about wildlife and the natural world.[16] This finding holds great promise. Perhaps we will witness considerable erosion in the value differences among blacks and whites as both groups learn to cultivate a better understanding of their mutual dependence and connection with the diversity of life.

Conclusion

This chapter has outlined the prevailing values toward nature and living diversity among Americans today. Some encouragement may be derived from the considerable affection and ethical concern found among many Americans toward familiar members of the animal world. Moreover, authoritarian and exploitative attitudes toward nature and wildlife have declined, though utilitarian perspectives still prevail in much of American society.

Despite these positive trends in perceptions of the natural world, most Americans remain fixed on a narrow segment of the biotic community—largely vertebrate animals, particularly creatures of special historical, cultural, and aesthetic significance. Concern for biological diversity and natural process continues to be limited and superficial. A person's willingness to grant species ethical standing or other positive values appears to depend on the presumption of the species' sentience, intelligence, and behavioral features reminiscent of human experience. Indifference, sometimes even hostility and fear, prevail toward much of the biological kingdom. Few people evince any scientific or ecological interest. Most Americans reveal limited knowledge or conceptual understanding of nature and biological process, except for creatures of practical or cultural importance. Only moderate interest is expressed toward direct contact with wildlife and the natural world. The great majority of Americans fail to appreciate the extent to which the intellectual quality, emotional value, and material well-being of their lives depend on an abundant, healthy, and diverse living world.

The United States constitutes, of course, a large and varied nation. Considerable diversity in wildlife and environmental values can be found among various demographic groups. Greater appreciation, concern, and interest are especially characteristic of younger and better-educated Americans. The positive appreciation of nature and wildlife seems to prevail among the college-educated, whatever their major, and is only slightly influenced by higher income and perhaps greater leisure time. Urban/rural, gender, and ethnic differences are substantial, as well, and, in varying ways, have important management implications. The following chapters will draw on these demographic comparisons as we consider the influence of varying activities, species, and cultures on human values of nature and living diversity.

CHAPTER 4

Activities

ALL PEOPLE EXPERIENCE nature and living diversity every day, though many fail to recognize the myriad ways the natural world enters their lives. Even in the most densely populated city, people encounter a wide variety of plant and animal species in their streets, parks, homes, and even businesses. Children particularly revel in the opportunity to explore and engage their curiosity for the nonhuman world, and many adults take sustenance in various reminders of the natural world around them. The sobering reality of an increasingly urban society, however, is how much the experience of wildlife and nature seems indirect, secondary, and simply unrecognized.[1] Restricted open spaces, zoos, and the television set have become for many the primary means for engaging nature and living diversity, which appears to grow ever more peripheral and pushed to the edge of modern consciousness.

A number of Americans still seek direct contact with wildlife and nature, though often in recreational contexts. A small but declining proportion of Americans continue to derive their livelihoods from exploiting the natural world. For the majority of Americans, however, the vicarious experience of animals through zoos, film, television, and other indirect means remains the predominant basis for encountering nature and living diversity.

This chapter explores some of the characteristics and consequences of various activities involving animals and nature—particularly how these interactions affect people's values and knowledge of the natural world. The sample of activities explored here covers five kinds of interaction: consumptive, nonconsumptive, indirect, vicarious, and abusive relationships between people and animals. Consumptive activities include the direct harvesting and removal of creatures from their habitats resulting in their death. The ancient practice of hunting is the focus of this inquiry. For the nonconsumptive uses of nature we will focus on the recreational observation of wildlife, mainly birdwatching. For the indirect encounters with animals we will consider wildlife in captivity, mainly zoos. The vicarious experience of nature will be explored by focusing on film and television. The chapter concludes by considering the psychological and behavioral consequences of deliberately abusing animals and nature.

Hunting

Hunting today is primarily a recreational activity in America, though still replete for many with practical significance. Hunting is one of the oldest of human practices. Indeed, according to some it has been the primary basis for our species' social, intellectual, and evolutionary development.[2] Perhaps 99 percent of human tenure on earth has been as a hunter-gatherer, representing 60 percent of the estimated 150 billion people who have ever lived. Only during the past ten thousand years have people developed a significant capacity to domesticate plants and animals, harness energy beyond the human body, or engage in large-scale toolmaking. The industrial era of a few hundred years represents but a blink of the eye in human evolutionary time.

It should not be surprising, then, that many of the behavioral and emotional characteristics which distinguish the human animal developed as a consequence of the hunting-gathering experience—including much of our capacity for ingenuity, invention, rapid problem solving, memory, language, specialization, and fabrication. The hunting-gathering way of life proved so stable and so successful for such a long period of time that some have suggested it might still represent the most likely basis for the continuing survival of the human species.[3] Modern agriculture and industry, despite having created huge material surpluses and intellectual advances, have also created teeming human populations, the capacity to wage wars of massive destruction, widespread pollution, and an unprecedented destruction of biological diversity and natural process. The paradox of modern life is that our immense power and productive capacity have not necessarily advanced the survival prospects of the human species. The seemingly primitive hunter-gatherer way of life—a life of small populations, few fixed territories, and scant material possessions—might indeed promise a greater likelihood of human continuity than our current extraordinarily complex and contentious industrial-technological existence. The protection of the earth's few remnants of hunting-gathering culture may be important not as expressions of compassion for the powerless, or to preserve quaint museum pieces, but as a critical survival option for the human species.

The recent emergence of extensive antihunting sentiment constitutes a remarkable development given the evolutionary importance of hunting and gathering. It suggests a fundamental change in perception and understanding of human relationships with the natural world. More than any other kind of human/animal interaction, participation in hunting versus opposition to it represents a particularly sensitive barometer of conflicting attitudes toward animals and the natural world.[4] An examination of perspectives for and against hunting involves such issues as human exploitation of the natural world, the role of socially sanctioned violence in modern society, the causes of species endangerment and extinction, and the ethics of inflicting death for reasons of necessity and recreation. The debate for and against hunting confronts one of the great paradoxes of human existence: on the one hand, humans appear equipped with the capacity and will to consume animal flesh; on the other, they remain equally capable of subsisting solely on vegetative and fruit matter. The philosopher José Ortega y Gasset captured this paradox when he remarked: "Do not forget that man was once a beast. His carnivore's fangs and canine teeth are unimpeachable evidence of this. Of course, he was also a vegetarian . . . as his molars attest. Man, in fact, combines the two extreme conditions of the mammal, and therefore he goes through life vacillating between being a sheep and being a tiger. . . . Hunting [consists of] a confrontation between two systems of instincts."[5]

The hunting-gathering way of life as a primary means of survival has largely disappeared from the modern world. Today it is predominant only among a limited number of increasingly marginalized peoples scattered across a few isolated deserts, high-latitude areas, and rain forests. Recreational hunting, however, remains a widespread activity, particularly in the United States. Some understanding of the significance of recreational hunting in America may be gleaned by reviewing its historical development since the colonial period.

During the seventeenth and eighteenth centuries, hunting was almost universally accepted by the European settler and most indigenous peoples. Although the taking of game in colonial America often represented a necessity, it also symbolized a freedom that had been largely denied in the Old World where wildlife was frequently treated as the property of a landed elite.[6] The European settler in North America reveled in the profusion of game animals—and in the right to possess weapons, which had been largely denied by the Old World's ruling classes. Unrestricted hunting in America, however, particularly with the development of modern weaponry, commerce, and transportation, resulted in an extraordinary and unsustainable waste of wildlife. Many species—including various furbearing mammals, birds of the forests and plains, marine creatures, large predators, and others—became extinct. The intensity of the slaughter is still hard to imagine. John James Audubon offers some perspective in describing an 1827 hunt of the passenger pigeon, an animal once capable of blackening the daytime sky with its remarkable numbers, yet eventually driven to extinction:

> Few pigeons were to be seen before sunset; but a great number of persons, with horse and wagons, guns and ammunition, had already established encampments on the borders. . . . Suddenly, there burst forth a general cry of "Here they come!" The noise which they made, though yet distant, reminded me of a hard gale at sea, passing through the rigging of a close-reefed vessel. As the birds arrived, and passed over me, I felt a current of air that surprised me. Thousands were soon knocked down by polemen.

> The current of birds, however, still kept increasing. The fires were lighted, and a most magnificent, as well as wonderful and terrifying sight, presented itself. The Pigeons, coming in by thousands, alighted everywhere, one above another, until solid masses, as large as hogsheads, were formed on every tree, in all directions. Here and there the perches gave way under the weight with a crash, and, falling to the ground, destroyed hundreds of the birds beneath, forcing down the dense groups with which every stick was loaded. It was a scene of uproar and confusion. I found it quite useless to speak, or even to shout, to those persons who were nearest to me. The reports, even of the nearest guns, were seldom heard. . . . No one dared venture within the line of devastation; the hogs had been penned up in due time, the picking up of the dead and wounded being left for the next morning's employment. . . . The uproar continued . . . the whole night. . . . Toward the approach of day, the noise rather subsided. . . . The howlings of the wolves now reached our ears; and the foxes, lynxes, cougars, bears, raccoons, opossums, and pole-cats were seen sneaking off from the spot, whilst eagles and hawks, of different species, accompanied by a crowd of vultures, came to supplant them, and enjoy their share of the spoil. It was then that the authors of all this devastation began their entry amongst the dead, the dying, and the mangled. The pigeons were picked up and piled in heaps, until each had as many as he could possibly dispose of, when the hogs were let loose to feed on the remainder.[7]

Despite the devastating effects of market hunting on America's wildlife, it was hunters, albeit the recreational sort, who eventually helped create the basis for modern wildlife conservation and protection.[8] Sport hunting emerged in America toward the latter half of the nineteenth century along with the country's newfound wealth. This new type of hunter focused mainly on the recreational pleasures of the activity, denying that wildlife's value derived only from its economic and material benefits. Sport hunters promoted controls over hunting not unlike those that had been reviled in the Old World. Like the European hunter, the American sportsman appeared motivated by the shift from wildlife abundance to scarcity, as well as a class bias against those who offensively harvested big game, songbirds, and small animals for the pot and the market. The American sport hunter had, in many respects, more in common with the European gentleman hunter than with his frontier countrymen.

Sport hunters eventually helped to establish an American wildlife management system based on eliminating the commercial harvest of wildlife. Game limits were instituted; hunting licenses were required; even an unprecedented police force was created to enforce the newly established wildlife laws. The new management system became a great success. Indeed, many previously depleted species, such as white-tailed deer, eventually returned to population levels comparable to those encountered during the colonial period. Recreational hunting flourished, too, despite licensing requirements and special taxes.

Today hunting constitutes the pastime of some 15 to 20 million Americans.[9] The proportion of Americans who hunt, however, has been steadily declining for a quarter of a century while the ranks of those opposed or indifferent to the activity have swelled. This change signifies a fundamental shift in the cultural context of hunting in America. Hunting participation historically depended on recruitment from a rural-based, extended-family network, as fathers, grandfathers, and other male role models socialized young boys from one generation to another, often within stable territorial boundaries.[10] The decline of rural living, the disintegration of the extended family, the increasing transience and mobility of Americans—all have altered the social conditions at the core of hunting's continuity and succession. American attitudes toward nature have shifted as well—particularly, increasing objection to the recreational killing of animals.

Although the proportion and, in recent years, absolute number of hunters have declined, the number of recreational fishers has significantly increased. This change may simply reflect greater opportunities for fishing than hunting in modern America, particularly in an increasingly urbanized society. Attitudes, however, have also played a role. Most people view fish as lacking the capacity for pain and suffering, while a majority regard large land mammals and birds as both sentient and intelligent creatures. Even the distinction between hunting and fishing reflects the tendency not to recognize fishing as just another form of human predation of animals for food or sport.

Opposition to hunting has dramatically increased during the twentieth century, particularly during the past fifty years.[11] As Figure 11 suggests, most Americans oppose hunting for strictly recreational purposes—and even larger numbers object to trophy hunting. Almost all Americans, however, support hunting to harvest the meat, whether the hunter is a Native American or not. A majority of Americans continue to approve of hunting if the purpose includes both recreation and use of the meat. Most hunters claim both motivations for participating in the activity.

The large number of antihunters contrasts with the nearly universal acceptance of the activity prior to the eighteenth century. Widespread opposition to hunting first prominently emerged during the eighteenth and nineteenth centuries in connection with the humane movements in urban England and America.[12] Initially concerned with cruelty to domesticated animals and child labor, this humane focus eventually extended to the presumed suffering of zoo animals, wildlife used for entertainment purposes, and hunting. Recreational hunting was considered particularly objectionable, although the presumed cruelty of the leghold trap also aroused considerable ire. We will undertake a more detailed examination of opposition to hunting later, but first I want to review some of the reasons why people hunt in America today.

The prohunting sentiments of the pioneering ecologist Aldo Leopold illustrate one motivational aspect of modern hunting. An ardent sportsman for much of his life, Leopold attributed his appreciation and even understanding of wildlife and ecology to his early experience of the land and its creatures through hunting. The young Leopold, nonetheless, often hunted for less complicated reasons, including the challenge, adventure, and desire for conquest offered by the sport.[13] Eventually he came to view hunting as a vehicle for achieving a deep participatory involvement in nature, including a profound understanding of the human connection to biological systems. He regarded hunting as a vivid reminder of human historic and evolutionary connections with biological reality. Paul Shepard describes this "split rail" value of hunting: "What does the hunt actually do for the hunter? It confirms his continuity with the dynamic life of animal populations, his role in the complicated cycle of elements . . . and in the patterns of the flow of energy. . . . Aldo Leopold postulated a 'split rail value' for hunting, a reenactment of past conditions when our contact with the natural environment and the virtues of this contact were less obscured by the conditions of modern urban life. . . . Regardless of technological advance, man remains part of and dependent on nature."[14]

Leopold denounced the commercialization of much modern recreational hunting, where the value of the activity all too often seems measured by the number of hunters and the amount of game bagged. Based

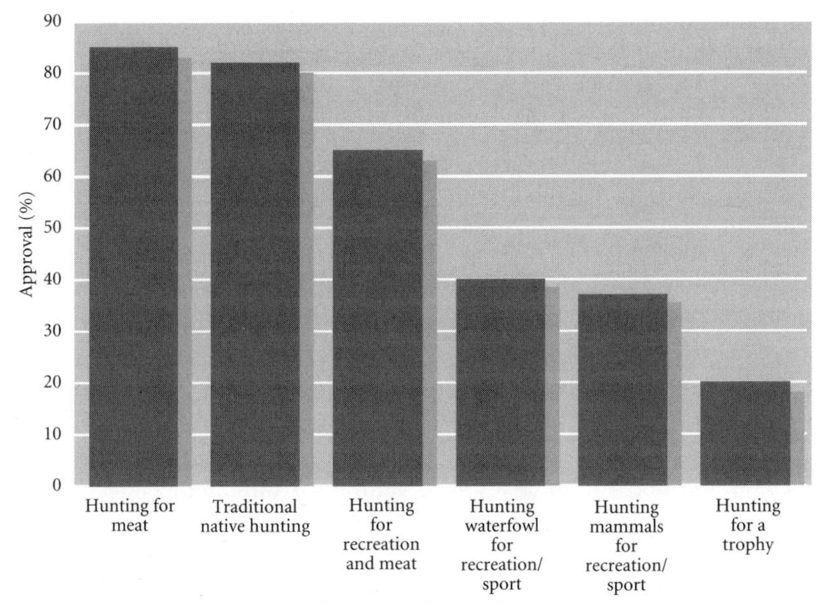

Figure 11. American Attitudes Toward Hunting

on our research, the kind of "nature hunting" advocated by Leopold comprises perhaps 10 to 20 percent of all hunters. For this minority of hunters, ecologistic and naturalistic attitudes constitute especially pronounced reasons for hunting. These nature hunters seek an active role in natural surroundings, and hunting represents a compelling opportunity for pursuing this interest. Exercising their role as predator offers nature hunters a chance for intimate experience of the complexity of ecological relationships and dependencies within a natural context. To the nature hunter, deep participatory involvement fosters a vivid appreciation and awareness of nature's many details and processes. As Ortega y Gasset relates: "When one is [nature] hunting, the air has another, more exquisite feel as it glides over the skin or enters the lungs; the rocks acquire a more expressive physiognomy, and the vegetation becomes loaded with meaning. All this is due to the fact that the [nature] hunter, while he advances or waits crouching, feels tied through the earth to the animal he pursues."[15]

Nature hunters further claim that the activity offers a heightened awareness and intuitive recognition of natural landscapes as a coherent and integrated whole. Most nature hunters argue that this ecologistic and naturalistic insight is difficult to achieve if "being in nature" is restricted to the role of observer or transient outsider. For them the pursuit of prey provides an unrivaled opportunity for entering into the life of another creature and partaking in its world. As John Madson and Ed Kozicky suggest: "The [nature] hunter deeply respects and admires the creatures he hunts. This is the mysterious, ancient contradiction of the . . . hunter's character—that he can at once hunt the thing he loves. . . . Part of the hunter's deep attachment to wildlife may stem from the fact that he sees wild creatures at their best—when they are being hunted. It is then they are strongest, freest and sharpest."[16]

For the nature hunter, the kill, while a necessary part of the activity, is not its primary focus. It is simply the reason and organizing basis for deep immersion and intimacy with natural surroundings. The paradox of rendering lifeless the object of appreciation and even affection has been rationalized by Ortega y Gasset's suggestion that the hunter "does not hunt in order to kill [but] kills in order to have hunted."[17] The nature hunter claims that the motivational core of the activity is not the successful kill but the involvement in natural process and an associated respect for the prey and its habitat. Through confronting death, the nature hunter asserts that he gains a better comprehension of life's precious and precarious meaning. As C. H. D. Clarke argues, "it's death that makes the spark of life glow most brightly measure for measure." A nature hunter elaborates: "Death acts as a reminder to us of what we need accomplish, of what goals and values . . . we should have for our lives. And wildlife and hunting are part of the reminder."[18] The nature hunter claims that a greater awareness of life's transient character forces a seriousness and enhanced respect for the precious gift of existence.

Nature hunting, however, is seldom the primary motive among most hunters. A far more common reason for hunting is to obtain the meat. Meat hunting accounts for four out of every ten hunters. Naturalistic and ecologistic values are not especially prevalent among meat hunters; these values are overshadowed by a far more pronounced utilitarian emphasis on nature and wildlife. This stress on killing wildlife for their practical value may seem surprising given how few Americans actually depend on wild game for survival or as a major food source. Yet a large number of rural Americans continue to kill game animals as an important protein supplement, particularly during the winter, not very different in motivation and purpose than raising farm animals or exploiting other wildlife for their fur, skin, and hides. Hunting for meat may seem anachronistic in modern society. But for many rural poor, this activity continues to be pragmatically significant, as well as an integral part of a lifestyle that focuses on extracting some portion of one's living from the land and its creatures.

For meat hunters, game animals primarily represent objects of practical significance. Unlike the nature hunter, the overriding motivation is the kill. The meat hunter does not express indifference or lack of affection for animals, but these considerations are typically subordinated to the higher priority of wildlife's usefulness or material significance. For most meat hunters, harvesting game constitutes the extension of an agricultural metaphor to wild animals. Wildlife should be harvested like any farm animal or agricultural commodity. As one meat hunter remarks: "If we can't eat the animal, then I will not kill it. I believe deer have to be harvested. I believe it's a harvestable crop. And if it's used for that purpose, it's not a whole lot different than going into the field and harvesting apples every year. You cultivate animals for what their purpose is, the same as cultivating any crop." For these hunters, wild animals represent a renewable resource managed for annual yields on a presumably sustainable basis. The overriding issue is not the legitimacy of this exploitation, but how much surplus can be efficiently and sustainably harvested.

A third kind of hunter, the sport hunter, is motivated less by naturalistic and utilitarian values, but more by the social, competitive, and sporting attractions of the hunt. Sport hunters account for perhaps one-third of all those who participate in the activity. Dominionistic attitudes are most pronounced among sport hunters, who relish the chance for exercising skill and asserting mastery over a presumably worthy opponent. The social benefits constitute another important motivation, especially the opportunities for male hunters to bond with other men.

Unlike the nature hunter, the sport hunter rarely evidences any unusual interest or knowledge of wildlife—or, like the meat hunter, much concern for the harvest value of wild animals. The dominionistic sport hunter covets most the opportunities for competition, conquest, and the exercise of dominance afforded by the hunting experience. Game animals signify an object of success and achievement rather than a subject of affection, intellectual curiosity, or practical value. Hunting represents more a human than animal-oriented activity, pursued mainly for its social rather than nature-related benefits. For most sport hunters, hunting provides valued opportunities for competition, camaraderie, and challenge. Lacking these qualities, the activity has little meaning or attraction. Vance Bourjailly has described this kind of sport hunting: "An advertisement of skill and a measure of success. [Wild game] are not items of food but trophies, something to get and display to fellow hunters."[19] A prizewinning trophy, such as a large head or fur pelt, rarely is essential, but the taking and displaying of game is thought to attest to the exercise of skill, courage, and fortitude in confronting a presumably cunning and formidable opponent.

Critics of sport hunting have especially derided its emphasis on dominance and masculinity. For most sport hunters, this focus represents less a demonstration of sexual virility than pleasure taken in the revelation of strength, power, skill, endurance, and, in effect, maleness. As one sport hunter remarked: "An important drive is the maleness of it. I don't know if the gun is a phallic symbol or not, but it's certainly a symbol of maleness, machismo, strength. Cats kill birds not because they want the bird. They kill it to confirm their prowess and ability. I think they share the same exultation in their prowess as the hunter." The sport hunter is motivated by the desire for experiencing the excitement of competition and challenge, for demonstrating physical dominance, and for having been tested and proved worthy. The activity's stress on masculinity is thought to signify admirable qualities of courage and hardiness.

But the hunters have their opponents. The widespread emergence of antihunting sentiment represents a relatively recent phenomenon in Western society. It first became prominent during the eighteenth and nineteenth centuries in northern Europe and America and grew especially after World War II. Antihunting views were closely allied with the humane movement, particularly the focus on cruelty toward domestic animals and child abuse at a time of rapid urbanization and industrialization. Antihunting sentiment also emerged along with the feminist and peace movements. Most hunters are male; some 80 percent of antihunters are female. Correspondingly, most hunting-related organizations consist of men, whereas the great majority of humane and animal welfare organizations' members are women.

Most antihunters oppose the activity because they believe it ethically wrong to kill animals for sport and object to the presumed pain and suffering inflicted on the hunted animal.[20] Little evidence supports the notion that most people oppose hunting because of opposition to guns or the presumably offensive behavior of "slob" hunters. Two distinct groups of antihunters emerge through our research. The larger group objects to hunting because of strong affection for individual animals—projected to include certain wild animals, particularly large mammals and birds. This group is especially distinguished by its pronounced humanistic value of nature and animals. Humanistic antihunters view as incomprehensible the taking of pleasure from killing animals for sport, which they regard as inflicting pain, anguish, and terror on creatures who possess a will to live not unlike that of a pet or, for that matter, other humans.

Humanistic antihunters focus on the presumed experience of the individual animal. In their view, species population levels or habitat maintenance are irrelevant in defending hunting. The notion of wildlife sometimes needing to be hunted for the "good" of the species or health of the habitat constitutes an abstraction of little meaning to most humanistic antihunters. Hunting seems repugnant for what it imposes on the creature itself, regarded as a victim not unlike humans if placed in similar circumstances. Given this emphasis on the presumed experience of the individual animal, many humanistic antihunters object to hunting large vertebrates (deer, ducks, bears, elephants) but do not oppose hunting snakes, fish, or invertebrates, animals typically viewed as lacking the capacity for pain and suffering.

Recreational hunting strikes most humanistic antihunters as taking intrinsic pleasure from the act of killing. Hunting often seems a deliberate act of destruction lacking adequate justification. Most humanistic antihunters are bewildered why anyone would consciously choose to eliminate a sentient life. As one antihunter remarked: "I don't see how anybody can get a thrill out of pulling a trigger and watching an animal fall over with blood spurting out of it. How anyone can take delight in reducing the wonder of any animal to a bloody mass of fur or feathers is beyond my comprehension. Why kill deer for sport? I don't go to the shelter to shoot cats and dogs when I want to have some fun. A deer, of all animals. It harms nobody. It's something that I can just look at and love."

A more philosophical perspective emerges among a smaller number of antihunters, distinguished most by pronounced moralistic values of nature and animals. Moralistic antihunters view this activity as one among many human/animal interactions they find objectionable for largely ethical reasons. Often they link their moral rejection of hunting to broader assumptions about proper conduct among people and society. Moralistic antihunters view hunting as ethically repugnant and even evil, not just the cause for presumed animal pain and suffering. This philosophical opposition to hunting is reflected in the views of Joseph Wood Krutch: "Killing for sport is the perfect type of that pure evil for which metaphysicians have sometimes sought. . . . Most wicked deeds are done because the doer proposes some good to himself. The liar lies to gain some end; the swindler and burglar want things which, if honestly got, might be good in themselves. . . . The killer for sport, however, has no such comprehensible motive. He prefers death to life, darkness to light."[21]

Moralistic antihunters most object to hunting because of its presumably degenerative effects on people. Beyond the death and suffering, hunting strikes many as an intrinsically degraded activity—debasing the hunter as much as destroying the victim. By deliberately killing another creature for sport, lacking a compelling justification, the hunter seems morally tainted and culpable, diminished by his or her willing participation in destroying another life. Many moralistic antihunters further condemn the society that condones this activity, believing that its acceptance fosters violent and antisocial behavior. The views of Albert Schweitzer are often cited in defense of this position, particularly his urging for a gentler, more reverent, attitude toward all living beings.[22] As one moralistic antihunter remarked: "I think essentially what you come down to is a Schweitzer approach, a kind of reverence for life. Hunting cheapens life, and it cheapens the perpetrator. It seems to me that whoever or whatever created this life put the same spark of life in a cockroach as he put in you and I. So the spark of life is just as valuable in that creature as it is in you and I. You really have no right to hunt and kill unless in self-defense or to survive."

Moralistic antihunters view life as precious and sacred, to be preserved except in highly unusual circumstances or absolute necessity. The nonessential pursuit of death through hunting seems vile. Deliberate killing for sport and amusement is thought to be the essential ethical difference distinguishing the hunter from the meat-eating nonhunter. As

Paul Breer observes: "Most people who eat meat are but dimly aware of what took place in the slaughterhouse. Eating meat is pleasurable only to the extent that it remains divorced from the actual act of killing. . . . Not so with the hunter."[23] Reversing Ortega y Gasset's notion of the hunter killing to hunt rather than hunting to kill, Breer suggests: "The hunter doesn't go hunting in spite of the fact that an animal will be killed; he goes hunting in order to kill the animal." Also, contradicting Ortega y Gasset, Breer insists that any form of hunting precludes the possibility of meaningfully entering into the life of another creature: "Being a part of nature implies feeling kinship with all nonhuman forms of life. It is impossible to feel this way if your purpose is essentially destructive. Hunting prevents one from cultivating a sense of belonging in nature. When man enters the woods as a predator he forgoes any chance he might have had to share in the lives of the many creatures that live there."

The ubiquity of death in nature strikes most moralistic antihunters as a flimsy excuse for inflicting even more unwarranted death and presumed suffering. Most view the hunted animal as possessing a will to live as profound as any human's. The moralistic antihunter asserts that ethical principles of empathy and kinship dictate extending moral rights to other creatures. Hunting, by definition, violates these rights.

Moralistic antihunting typically represents an integral part of a general philosophical perspective of how humans ought to treat one another. Moralistic antihunters object to many forms of animal exploitation: rodeos, trapping, fishing, laboratory experimentation, the consumption of meat, and, for a small number, the use of any animal product derived from dead or live creatures. Sympathy for the exploited is a critical element in a perspective that often embraces a pacifistic philosophy. Hunting seems to be a form of violent antisocial conduct; the hunter is portrayed as sadistic; the activity is deemed likely to foster psychopathic tendencies. From this high moral plane, antihunting sentiment stems less from sympathy for the animal than from a fervent belief in the activity's intrinsically degenerative impact on people and society.

Commercial Hunting

No discussion of hunting in America would be complete if it failed to consider the subject of commercial hunting. The American legal approach to hunting is based largely on denying commercial markets to wildlife. Two important exceptions, however, include fur trapping and commercial fishing—activities not usually viewed as forms of hunting but nonetheless involving the pursuit and killing of wild animals. People occasionally trap animals for recreational purposes, but most trappers appear motivated by the opportunity for marketing animal pelts, hides, and other body parts used mainly for clothing and adornment. Commercial fishing, in many respects, can be considered the last great form of market hunting left in the modern world.

Considerable controversy has developed around the trapping of wild animals, especially the trapping methods employed and the ethics of killing animals for luxury purposes.[24] The use of the leg-hold trap particularly outrages many, including those opposed to recreational hunting. Many regard trapping as intrinsically cruel and the cause of great physical pain and psychological terror for the animals.

Far less ethical objection has been directed at commercial fishing—partially because of prevailing assumptions regarding the lack of sentience and individuality among fish. Nevertheless, considerable opposition has emerged to fishing practices resulting in extensive mortality to marine mammals, particularly whales and dolphins. Increasing ethical concern has also been directed at modern fishing practices that overexploit certain fish species, such as the bluefin tuna, to the point of commercial and near biological extinction.[25] Although principles of sustainability and rational management have often been trumpeted, these goals have seldom been achieved in the face of competitive world markets, modern fishing technology, and uneducated affluence.

Trappers and commercial fishers in our studies reveal unusual patterns of values toward nature and living diversity. As might be expected, both groups voice strong support for utilitarian values and the right of humans to exercise dominion over the natural world. They also express few moralistic reservations regarding the exploitation and treatment of commercially valuable species. Yet trappers and fishers alike reveal considerable knowledge, affection, and interest in wildlife and the outdoors and express strong support for protecting natural habitats. This somewhat unusual pattern of naturalistic interest and humanistic detachment, as well as ecologistic concern and moralistic indifference, appears characteristic of people who live close to nature but derive much of their income from exploiting and extracting wild living resources.

Birding

Watching wild animals constitutes an ancient human interaction with the natural world. What distinguishes the activity today is its organized recreational development. Wildlife observational interest has expanded greatly during the twentieth century. Today it encompasses a wider range of species, from birds to whales, and recent years have even witnessed an emerging recreational interest in viewing butterflies and other invertebrates.[26] As one expression of this burgeoning phenomenon, nature and wildlife tourism may now account for as much as 10 percent of the $300 billion world tourism market, growing at an estimated 10 to 20 percent rate in recent years.[27] America's national parks now accommodate some 300 million visitors annually, many drawn by the opportunity to see wild animals; in Kenya alone, wildlife viewing has become the single biggest contributor to that country's foreign exchange earnings. This remarkable growth in wildlife viewing has also extended to the marine realm. Whalewatching did not exist as an organized activity a half century ago, yet today it accounts for some 3 million participants in thirty countries generating almost a half billion dollars annually.[28] Moreover, the explosive growth of wildlife viewing has occurred alongside a significant decline in hunting and other forms of consumptive wildlife use.

The number of people who participate in wildlife viewing and other forms of nonconsumptive use is hard to determine. A fundamental problem stems from the difficulty of actually defining the activity. Observing wildlife can encompass a bewildering assortment of activities involving varying levels of commitment and expression. Birdwatching can range, for example, from looking at pigeons in a city park to taking a casual stroll with binoculars in a nearby wetlands to committing considerable time and expense pursuing rare birds in distant lands with the most sophisticated equipment. Hunting wildlife, by contrast, involves distinctive behavior and the use of specialized equipment, resulting in a clear impact on the targeted animal.

Reliable statistics regarding wildlife viewing in America are not easy to obtain, although estimates by the U.S. Fish and Wildlife Service suggest a majority of Americans annually engage in some form of wildlife observation.[29] Most of this activity is casual, local, and rarely requires much investment of time, energy, resources, or skill. The number of people who actively participate in wildlife viewing with a commitment similar to hunters appears relatively restricted. Perhaps 5 to 10 percent of the American population might be called active birdwatchers, for example, insofar as regularly using binoculars, consulting field guides, being able to identify forty or more species, and traveling at least some distance for the specific purpose of observing birds in the wild. This figure represents roughly two-thirds of the number of hunters in America.

Birding illustrates some of the characteristics and values associated with wildlife observation. Birds have long fascinated people—indeed, perhaps more than any other taxa in the animal kingdom. Studies in the United States, Germany, and Japan have revealed birds to be the most popular class of animals.[30] Given our own status as mammals, this finding may surprise some. Certain characteristics of most birds, however, tend to make them particularly appealing, including their colorful qualities, their capacity for flight, and their being diurnal. The majority of mammals, by contrast, lack color, cannot fly, and are nocturnal. The capacity for flight has also allowed many birds to remain visually accessible even in areas of dense human habitation.

The visibility and aesthetic appeal of birds have fostered the tradition of recreationally observing these creatures, an activity that developed particularly in northern Europe and the United States. Birdwatching never became especially popular in southern Europe, although shooting songbirds for food has long been an established practice in this area, much to the chagrin of many northern Europeans.[31] Birds were observed, identified, and even painted during the European settlement of North America, most prominently associated with the works of John James Audubon and Alexander Wilson. As an organized recreational activity, birdwatching emerged conspicuously during the latter half of the nineteenth century with the formation of the Audubon societies, the first established in Massachusetts in 1886. The Audubon societies particularly coalesced around opposition to the bird trade and, along with sportsmen and animal welfare groups, helped create the basis for modern American conservation. These three constituencies became especially outraged by the mass destruction of plume birds for the millinery and hat markets, which had pushed many species to the brink of extinction.[32] Their combined efforts led to the passage of America's first federal wildlife legislation, the Lacey Act of 1900, which undermined the commercial trade in wildlife, established a basis for national wildlife law, and assisted in the creation of the national wildlife refuge system. Above all, the Audubon societies and their allies laid the foundation for the American environmental movement, which eventually became a major force in developing this country's system of natural resource law and regulation.

People obviously watch birds for many reasons. The casual birder may simply enjoy seeing birds in a mainly residential context without having to exercise much effort or skill. For the casual birder, the aesthetic qualities and emotional appeal of birds often represent the activity's greatest attraction. For a smaller group of committed birders, this activity often reflects considerable interest in direct contact with nature and living diversity, in seeing as many species as possible, as well as occasionally exercising a more ecological fascination in understanding birds and their habitats. These active birders often employ specialized equipment, travel long distances, and possess considerable knowledge of animal life.

Active and committed birders in our studies are usually distinguished by an unusual degree of appreciation, interest, knowledge, and concern for nature and biological diversity—perhaps to a greater degree than we encountered among any other animal-related group with the possible exception of nature hunters. As the results of Figure 12 suggest, active birders evince considerable interest and understanding of wildlife. Their ecologistic and knowledge scale scores are, in fact, higher than any other wildlife activity group we examined. Active birders also express strong interest in direct contact with wildlife and a consistent willingness to protect animals and their habitats even when this means considerable economic sacrifice. This conservation commitment rarely involves a moralistic rationale, however, as active birders attained only moderate scores on this scale and do not reveal any unusual opposition to hunting or other consumptive wildlife use. Nor do active birders express any greater affection for individual animals than does the general population. Active birders appear, therefore, to articulate their support for wildlife protection in mainly ecologistic rather than moralistic or humanistic terms. Appreciation of wildlife among active birders often seems more intellectual than passionate, more generalized than focused on the creature itself.

Various characteristics of birdwatching may explain this configuration of interest, knowledge, and concern for wildlife and natural habitats. First, observing birds typically focuses on species rather than individual animals: most people view the individual bird as just a representative of its kind. This emphasis on species, rather than individual animals, may foster an ecologistic rather than humanistic attitude and thus a related stress on the well-being of the group rather than the welfare of the single creature. Second, birdwatching tends to direct one's attention to certain landscapes and environments. Many bird species are specialists who closely associate with particular habitats as sources of food, cover, and nesting areas. This characteristic may cause birdwatchers to become more aware of the ecological relationships among varying species and their habitats. Finally, many birds possess high metabolic rates that, combined with habitat dependence and specialization, make these creatures especially vulnerable to environmental disturbance. This feature may sharpen concern among active birders when they see the widespread damage inflicted by people on many natural environments.

These various results suggest that active birding encourages sensitivity, understanding, and appreciation of the value of living diversity and its conservation. From an educational perspective, this finding may be important, for birding is an easy activity to promote compared with many other forms of wildlife observation and use. Birds can be viewed easily in most settings, even urban ones, including parks, wetlands, golf courses, and other open spaces. Unlike hunting and fishing, birding requires no special licenses or permits, is not restricted to certain seasons or habitats, has no bag limits, and involves only a modest investment of time and resources.

When birding and other forms of wildlife observation are practiced with restraint, they offer a steady stream of diverse benefits with little diminution of the resource. Even so, birding is not an intrinsically benign or inconsequential activity. Birders and other wildlife observers can

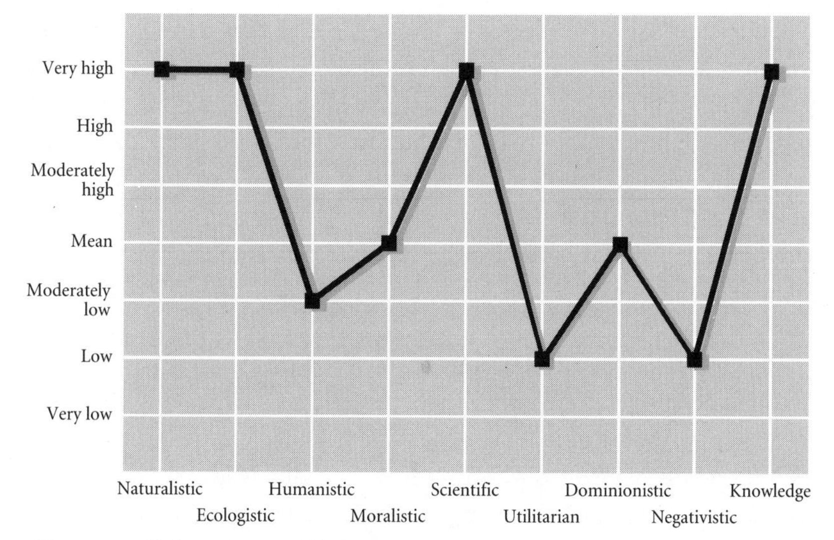

Figure 12. Values Among Birders

trample vegetation, damage habitat, cause soil erosion, disturb and harass animals, and even directly harm wildlife.[33] Some forms of wildlife observation require extensive infrastructural support—hotels, roads, motorized equipment, and large-scale development in remote areas—that can cause great harm to wildlife and natural habitats.

The interests of conservation have often been enhanced by the economic expenditures of growing numbers of nature tourists attracted to the wonder and beauty of birds and other wildlife. Birders, ecotourists, and other wildlife observers, however, must recognize the importance of understanding and protecting the biological requirements of varying species and habitats, as well as the need for informed experience of nature and living diversity. All forms of wildlife use, consumptive or nonconsumptive, necessitate respect, restraint, and appreciation for the gift of nature temporarily placed in their trust.

Zoos

For many Americans, visiting zoological parks represents one of the few opportunities they may have for experiencing exotic wild animals. The popularity of zoos is reflected in the suggestion that more Americans visit a zoo annually than attend all professional football, baseball, and basketball games combined.[34] An estimated 125 million Americans attend the country's more than one hundred zoological parks each year, and more than 350 million people visit the world's six hundred plus zoos. This extraordinary popularity may reflect the continuing desire of most people, even in modern urban environments, to experience some of association with living diversity, no matter how fleeting and superficial.

Despite its popularity, tensions characterize the modern zoo. From a positive perspective, zoos offer many Americans, particularly city residents, a rare opportunity for experiencing wildlife. A measure of authenticity can be obtained from confronting the living wild creature that no amount of reading, film viewing, or other vicarious experience can replace.[35] The zoo visit may encourage appreciation, knowledge, and even awe inspired by life's remarkable diversity.[36] Scientific studies of zoo animals have led to many important findings, including information on species that would be difficult, sometimes impossible, to collect in the wild. Captive zoo animals can further provide genetic reservoirs for endangered species, sometimes preventing their extinction and offering the possibility of their future restoration to the wild.

From a more skeptical perspective, the viewing of zoo animals can profoundly distort the typical visitor's attitudes toward nature and wildlife.[37] The spectacle of confronting incarcerated animals in highly artificial environments who are forced to live in close proximity to one another may fundamentally compromise the ability of these creatures to instruct the visitor with naturalness and grace. Accidental injuries, deaths, and considerable psychological distress may result from the capture and incarceration of zoo animals. Isak Dinesen depicts this tension, and the possible tragedy of the zoo animal, in speculating on the experience of two giraffes on their way to a life of captivity:

> Upon the deck there stood a tall wooden case, and above the edge of the case rose the heads of two Giraffes. They were . . . on board the boat . . . and going . . . to a travelling menagerie. The Giraffes turned their delicate heads from the one side to the other, as if they were surprised. . . . The world had suddenly shrunk, changed and closed round them. They could

> not know or imagine the degradation to which they were sailing. For they were proud and innocent creatures, gentle amblers of the great plains; they had not the least knowledge of captivity, cold, stench, smoke, and mange, nor of the terrible boredom in a world in which nothing is ever happening. Crowds . . . will be coming in from the wind and sleet of the streets to gaze on the Giraffes, and to realize man's superiority over the dumb world. . .
>
> In the long years before them, will the Giraffes sometimes dream of their lost country? Where are they now, where have they gone to, the grass and the thorn-trees, the rivers and water-holes and the blue mountains? The high sweet air over the plains has lifted and withdrawn. . . . As to us, we shall have to find someone badly transgressing against us, before we can in decency ask the Giraffes to forgive us our transgressions against them.[38]

As Dinesen intimates, the zoological park seems caught, on the one hand, between the goals of mass entertainment and inexpensive escape and, on the other, public education and conservation. The modern zoo appears to operate at the edge of potential: unsure of what it signifies to most people and uncertain of what it would like to become.[39]

This ambivalence is reflected in the historical origins of the zoo.[40] Wild animals have been displayed in captivity for thousands of years, and the zoo can trace its roots back to the royal menageries of ancient Greece, Rome, and Mesopotamia. The tradition established then, and still surprisingly prevalent, is one of incarcerating a wide assortment of exotic species largely for people's amusement. Zoos accessible to a wider public became established in Europe in the sixteenth and seventeenth centuries. The regime of Louis XIV created perhaps the first systematic collection of varying species, eventually becoming the prototype of the modern zoo. The zoological park in the United States developed during the nineteenth century. It became so popular that by 1950, nearly every major American city believed its cultural status somehow depended on possessing a zoological park. The modern zoo emerged in Germany during the early part of the twentieth century, though scientifically oriented zoos had been established in England in the late nineteenth century. Animal behavior, natural history, and the more humane display of animals began to rival convenience, entertainment, and amusement as the primary objectives of the zoo.

Until recently, only a handful of zoos insisted on naturalistic exhibits, public education, and wildlife conservation as major goals of the institution. The "best and brightest" zoos today claim to be centers of public learning and animal conservation. One prominent zoo director has remarked: "The ultimate justification of zoos as institutions . . . is the exhibition of animals for educational purposes."[41] The tension between mass entertainment and public learning remains considerable, however. Zoos are still uncertain about the ultimate compatibility of these two objectives, as well as the actual extent of learning experienced by the average zoo visitor.

A fundamental constraint on the zoo's educational impact derives from the structure of the typical visit. Most zoo visits are voluntary, informal, exploratory, and self-paced.[42] The average visitor seeks a pleasant encounter with animals that does not involve any strenuous learning. Most zoo visitors perceive education as difficult and dull; the conservation of wildlife, especially rare and endangered species, is considered gloomy and depressing. The typical zoo visitor seeks a safe family or group experience in an agreeable parklike setting. The zoo visit represents a chance to observe other people as much as see wildlife. Study and learning about animals may be an element of the experience, but as often as not it is a peripheral one.

Most zoos, responding to this reality, have allocated little staff, space, or prestige to the function of educating the zoo visitor. Many zoos proclaim the fundamental importance of their educational mission, but few devote much in the way of resources toward cultivating an informed, appreciative, and concerned public. Most zoos provide only cursory signs on animal biology, behavior, and population status, and these signs are typically read for thirty seconds or less, the information rarely shared with others.

The typical zoo visit constitutes, as suggested, a family or group experience in a presumably pleasant and safe urban park. As Susan Swensen observes: "The trip to the zoo is a unique social occasion shared as a group with family and friends. Visitors use the zoo experience as a means of strengthening [social] ties."[43] Observational studies suggest that visitor groups often engage in activities quite unrelated to the animals: conversation, observing others, entertainment, and the like. When visitors do focus on the animal, they prefer active wildlife, with humanlike attributes, usually amusing and exotic creatures. Their conversations often involve anthropomorphic comments about prominent physical features of the animals and their behavior. Most visitors also desire, when possible, direct contact with the animals such as feeding and touching.

According to our research, the typical zoo visitor possesses limited knowledge and appreciation of wildlife.[44] The average visitor reveals largely humanistic and aesthetic attitudes toward animals. Wildlife interest focuses on just a few favored creatures regarded as physically appealing, unusual, and entertaining. Rarely is the zoo visit valued as an opportunity for naturalistic appreciation, scientific discovery, or ecologistic understanding. Perhaps most discouraging, the typical zoo visit appears to exert only a slight influence on people's understanding of animal behavior, biology, or conservation. Visitors emerge only marginally better informed following the average zoo visit. Although some data suggest that visitors develop more emotional affinity for animals following a zoo visit, this appreciation is seldom based on increased understanding.

Most visitors express strong support for displaying animals in more humane and naturalistic ways. To their credit many zoos have devoted considerable resources to creating larger, more stimulating, and more authentic-looking natural habitat exhibits. Still, the average zoo visitor expects to see as many animals as possible both quickly and conveniently. Spacious, naturalistic, humane exhibits often conflict with these objectives, especially when wildlife seek cover and hiding places.

The typical zoo experience appears to reinforce attitudes of mastery over animals and nature. The incarceration of wildlife essentially for entertainment affirms for many visitors the assumption of human superiority over the natural world. Only the most humane and ecologically oriented exhibits significantly counter the implicit suggestion of human control over nonhuman life. Nevertheless, a small proportion of zoo visitors do evince pronounced interest, knowledge, and fascination for animals and the natural world. For this group, the modern zoo enhances their understanding and appreciation of wildlife, as well as reinforcing their concern for the plight of declining wild animal populations.

Moreover, zoos and exhibits vary widely in their effect on visitors' values and knowledge of wildlife and nature. Exhibits with an educational focus—involving animals displayed in vegetative environments, simulating natural conditions, and using multimedia information displays—appear to exert a particularly positive impact. These salutary effects are greatest for zoos with a high proportion of indigenous wildlife closely associated with representative ecosystems found in reasonable proximity to the facility. Visitors to these zoos reveal substantially greater knowledge, conservation interest, and ecological awareness than those who attend traditional zoological parks. These results suggest that carefully designed, informative, and naturalistic exhibits, even those with conservation themes, can substantially enhance visitor appreciation and understanding of wild animals and their conservation.

The current controversy over the captive display of marine mammals should be noted here. This issue reflects many of the complexities and ambivalencies associated with the zoo. Aquariums and marine parks, particularly those with marine mammals, have become remarkably popular.[45] For example, visits to the thirty-five leading aquariums with marine mammals exceed 10 million people annually. Most people who observe marine mammals in captivity report how much they enjoy seeing these animals; moreover, they claim their support for marine mammal conservation has been enhanced by the experience.[46] Considerable advances in knowledge of marine mammal biology and behavior, including information relevant to their conservation, have also emerged from studies of these animals in captivity.

Like zoological parks, however, opinions are split over the value of aquariums with captive marine mammals. Extensive opposition has emerged in recent years—especially regarding the number of injuries, fatalities, and psychological distress experienced by marine mammals during their capture, as well as the high rate of mortality and shortened life spans of these captive animals, particularly cetaceans. Some people have voiced strong ethical objections to the very notion of marine mammals in captivity, especially when these animals are used for entertainment and to perform tricks. Animal rights activists have illegally freed captive marine mammals, and the recent film *Free Willy* romanticized the release of a killer whale. Despite the visitors' obvious enjoyment of the shows, most indicate they support the captive display of marine mammals only if significant educational and scientific benefits result.

Once again we note a basic tension between the positive educational, scientific, and conservation benefits of the modern zoo and aquarium versus the traditional emphasis on entertainment and amusement. The popularity of zoos and aquariums underscores their potential for cultivating an enhanced appreciation and concern for wildlife and the natural world. These facilities offer extraordinary opportunities for people to affiliate with wild exotic creatures. As institutions built around the incarceration and often frivolous display of wildlife, however, zoos and aquariums can foster attitudes of separation and dominion over the nonhuman world. Often the experience of wildlife in zoos and aquariums is dominated by the artificiality and harshness of the exhibits. The spectacle of captive animals offers but a shallow reflection of wild reality, and the confinement of innocent creatures for human amusement may exert a degrading effect. Indeed, a pervasive atmosphere of control and superiority could actually bolster the illusion of human hegemony, leaving the average visitor more arrogant than ever toward the nonhuman world.

Zoos and aquariums that have committed effort and resources to the better display of their animals may transform these institutions into more positive opportunities for cultivating appreciation, understanding, and concern for living diversity. Much more, however, needs to be done. Most zoos have yet to dedicate themselves to education as the core of their mission. A much greater emphasis is needed on exhibiting all forms of animal life, invertebrates included, developing more ecologically oriented displays, and instituting better naturalistic exhibits involving plants, soils, and landscape features. Exhibits should be tied to human experience—including a stress on the importance of conserving living diversity as a basis for human well-being. Educational efforts should be more extensive and diverse—employing varying communication media and tailored to a wide diversity of zoo visitors. Zoo animals should be treated not as captives but as ambassadors from other worlds—temporarily on loan to instruct people of the inherent beauty and wonder of nature and living diversity.

Film and Television

Portraying wildlife and nature on film and television constitutes a revolutionary development offering people an unprecedented opportunity to enter into the lives of other creatures.[47] At one time this contact with the natural world was inaccessible to all but the most privileged, knowledgeable, and widely traveled. The widespread popularity of nature-related film and television in just a few short decades is attested to by the majority of Americans, Japanese, and Germans who report viewing at least one such film or television program during a typical year—and one-third of them say they regularly watch wildlife shows on television. Indeed, film and television may represent the most frequent form of contact between people and wildlife today.

The impact of wildlife film and television remains uncertain, however, given the vicarious, secondhand, and indirect character of the contact it offers with the natural world. Wildlife on film and television may constitute an anesthetized experience of nature fundamentally compromised by its occurence within the comforts and artificial confines of the human habitation. This encounter with wild lives is essentially contrived, divorced from the constraints, complexities, and realities of the real world. It remains uncertain how much this vicarious experience can positively shape people's values of nature and living diversity.

Despite the limitations of the visual media, research suggests that film and television may exert a major impact on people's perceptions of the natural world. One finds an intriguing association between the remarkable increase in public appreciation of wildlife during the past half century and the development of nature-related film and television programming. Moreover, most Americans and Japanese report that nature programs have greatly influenced their ideas and knowledge about animals and the natural world. These studies also suggest that the impact of television on basic wildlife values may be limited, however, and largely restricted to increasing general appreciation and affection for nature and animals. There seems to be very little intellectual change and minimal increase in ecological awareness or ethical concern for the natural world and its conservation. In short, the impact of television and film seems transient. Rarely does it involve profound changes in perception without repeated exposure and personal experience. A more significant effect has been observed in studies of popular feature-length films such as *Never Cry Wolf,* as well as detailed and complex television programs about animals and the natural environment. These viewing experiences appear to influence people's understanding and concern for protecting nature. Moreover, these programs seem to motivate people to seek greater recreational contact with wildlife and the natural environment.

During the early years of nature-related film and television shows, depictions of wildlife tended to be anthropomorphic, often using trained animals in highly contrived ways. These programs frequently emphasized entertainment along with substantial doses of adventure, action, and the "tooth and claw" battle of humans versus nature. Animals often displayed humanlike attributes—particularly those depicted in the so-called family-oriented films and television shows. Ecologically realistic portrayals of animals and natural habitats rarely appeared in the early days. Conservation themes were practically nonexistent.

Wildlife television programming has changed considerably during the past two decades. Scientifically sophisticated and problem-oriented depictions of wildlife, though still far from the norm, are more frequently encountered today. The popularity of animal-related television programming, however, has declined. Wildlife television shows tend to be confined to public broadcasting stations viewed by a minority of well-educated and affluent people. The popularity of these shows on public broadcasting is indicated by the fact that two-thirds of the most-watched public television programs have had a nature—particularly wildlife—theme. Public television programs rarely range much beyond natural history and wildlife from distant lands; few consider the links between people and nature or examine the modern biodiversity crisis. On network television, most nature shows continue to emphasize entertainment, adventure, and action, not ecology or conservation. These portrayals usually involve dramatic plots, beautiful scenery, and a good deal of sentimentality.

Considerable uncertainty remains about the impact of this revolutionary medium on human values of animals and the natural world. The vicarious experience of wildlife through film and television has clearly expanded consciousness and appreciation to an extent perhaps unrivaled in modern history. It is worth noting again the parallel development of this medium and expanded public concern for the natural world during the past half century. Doubt continues, however, regarding the lasting impact of this technology—particularly its capacity to communicate ecological complexity or the difficulties of nature conservation. Indeed, television's vicarious, indirect, and often unrealistic depiction of the natural world may promote naive and distorted assumptions about wildlife and its maintenance.

More innovative approaches might strengthen the positive impact of nature-related film and television programming. Since the public engages the visual media in a largely passive way, new approaches will be needed to fuse information with entertainment, ecological understanding with the desire for escape, and the conservation of living diversity with recreational pursuits. Unless a new paradigm of wildlife-related film and television develops, this revolutionary medium may never achieve its full potential for enhancing the public's understanding of living diversity and its importance to human life and experience.

Abusing Animals

This chapter has generally emphasized positive interactions between people and the living world. These varying relationships suggest how people's physical, intellectual, emotional, and spiritual well-being can be influenced by the quality of their experiences of nature and wildlife. Each activity noted here offers important opportunities for intimate and satisfying contact with the natural world.

The benefits we derive from rewarding experiences with nature suggest, by implication, that degraded or abusive relationships with animals might foster tendencies toward alienation, disaffection, and perhaps even harmful antisocial behavior. Does malicious conduct toward other creatures lead to similar actions toward people and society? We will consider this possibility by reviewing the results of research conducted by Dr. Alan Felthous and myself on the relationship between childhood animal abuse and later adult aggressive behavior toward people.[48]

The gravest perversion of the human/animal relationship may be the deliberate inflicting of pain and suffering on other creatures. Defining cruelty toward animals, however, depends on one's cultural and historical assumptions. Opinions differ sharply over whether bullfighting, cockfighting, trapping, or hunting constitutes cruelty toward animals. Perceptions of animal cruelty may also depend on the species in question. Most people limit such judgments to creatures that evidence the capacity for experiencing pain or suffering or are viewed as possessing a mental life. For most people, this means the higher vertebrates. Consequently, most people tolerate abusive behavior toward insects, spiders, fish, and even higher vertebrates with dubious reputations such as rats and rattlesnakes. Despite this cultural and taxonomic relativism, most Americans generally regard the deliberate and repeated infliction of harm, pain, and suffering on higher vertebrates as constituting cruelty toward animals.

The issue of animal cruelty emerged in Western society in the seventeenth, eighteenth, and nineteenth centuries, a time when rapid urbanization and industrialization led to extensive abuse of domestic animals. This concern was also influenced by emerging notions regarding the capacity of higher vertebrates to experience pain and suffering. Empathy for other creatures appeared to be fostered by the growing practice of bringing pets into the home as companion animals and making them quasi-family members. The historian Keith Thomas notes that pets "encouraged [people] to form optimistic conclusions about animal intelligence [and feeling] . . . stimulating the notion that animals could have character and individual personality; and . . . creating the psychological foundation for the view that some animals . . . were entitled to moral consideration."[49]

The idea also developed that cruelty toward animals could brutalize the human perpetrator, rendering such persons more likely to behave in similar ways toward other people. A popular nineteenth-century depiction of this possibility was Hogarth's *Four Stages of Cruelty:* a child tortures animals, beats horses during adolescence, and eventually kills a person as an adult.[50] The anecdotal association of childhood cruelty toward animals leading to violence in adulthood has continued to the present day.[51] Examples in the popular press include the "Son of Sam" murderer, who reportedly hated dogs and killed neighborhood animals; a California mass murderer with a history of childhood cruelties toward cats and dogs; a mass killer who immersed cats in battery acid as a child; the infamous Boston Strangler who trapped pets, placed them in orange crates, and shot arrows through the boxes; a young man who admitted killing people for fun and described how as a child he would put ammonia in fish tanks and watch the turtles turn white; the recent killer of twelve-year-old Polly Klaas who reported a history of repeated animal abuse in childhood.

The notion that cruelty to animals in childhood can lead to later adult violence against people has, however, rarely been demonstrated scientifically.[52] Although a few studies in the 1960s reported an association of childhood animal cruelty, fire-setting, and bed-wetting among adult violent criminals, these studies involved very small sample sizes, inadequately defined animal cruelty, and were marred by limited data collection techniques.

The study conducted by Alan Felthous and myself constitutes the most extensive research on the connection between childhood cruelty toward animals and later adult violent behavior. This investigation focused on repeated and significant acts of cruelty toward animals in childhood by persons with a clear history of aggressive and violent behavior as adults. One hundred and fifty men participated in the research: very aggressive and moderately aggressive criminals, nonaggressive criminals, and noncriminals. Since violent behavior may be an isolated act, criminals were classified as aggressive or nonaggressive based on a history of persistent violence toward other people rather than the reason for their incarceration.

Three hundred and seventy-three distinct acts of cruelty toward animals were recorded. These acts covered a wide range of behavior. Many involved behavior viewed by most people as heinous such as skinning animals alive, stoning and beating them to death, or severing body parts. Most of the cruel acts, however, were directed at creatures generally regarded as lacking in the capacity for pain and suffering, such as insects. Tearing the wings off bugs, for example, was cited by nearly one-third of the subjects, suggesting this behavior may be almost normative among young males in American society. Most people are disinclined to view callous behavior toward insects as indicative of cruelty toward animals.

Moreover, the great majority of reported cruelties toward animals happened only once in the person's childhood.

Repeated acts of childhood cruelty toward socially valued animals occurred to a far greater degree among aggressive than nonaggressive criminals and noncriminals. As Figure 13 reveals, one-quarter of the violent criminals reported five or more instances of childhood cruelty toward animals—compared to less than 6 percent of moderately and nonaggressive criminals, and no occurrence among noncriminals. A much higher percentage of cruel behavior toward animals among violent criminals also involved serious forms of animal abuse. Violent criminals, for example, far more often reported stoning and beating animals than either nonaggressive criminals or noncriminals. Moreover, violent criminals scored significantly higher than the other groups on a childhood aggression toward animals scale, which included abusive behavior not meeting the strict definition of animal cruelty.

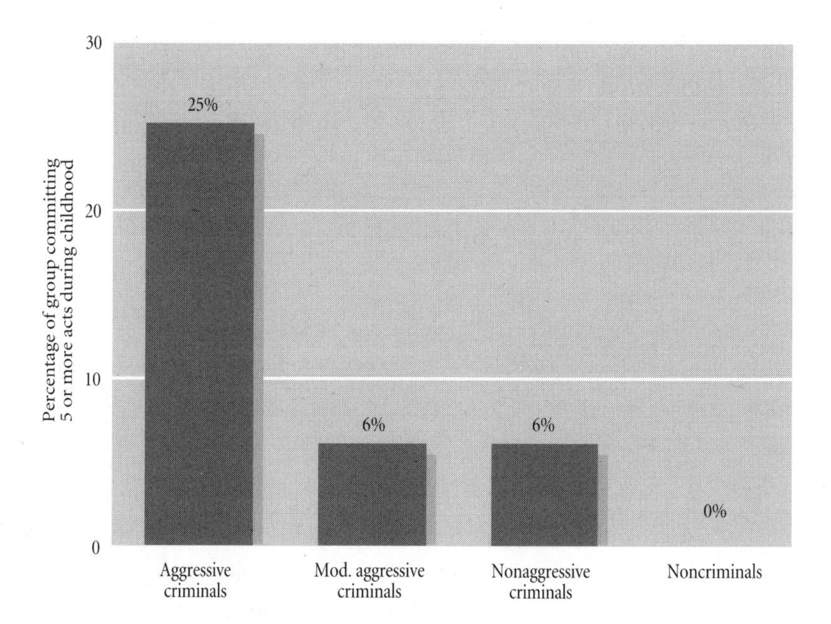

Figure 13. Childhood Cruelty Toward Animals

Our results clearly reveal a strong link between serious childhood abuse of animals and recurrent violence against other people in adulthood. There is also a close connection between animal abuse and violence in the families of violent criminals during childhood. Nearly three-quarters of violent criminals were raised in abusive families, for example, in contrast to a small minority of nonaggressive criminals or noncriminals. Nearly half the violent criminals also reported extensive alcohol and drug use among family members during childhood, in contrast to much smaller percentages among nonaggressive criminals and noncriminals.

The association of family violence in the childhood of aggressive criminals suggests a possible motivation for their cruelty toward animals. A child's inclination to inflict serious injury on animals may reflect displaced feelings of hurt, anger, and rage toward other people, especially adult family members. Perhaps animals serve as scapegoats for the deep hurt that would be nearly impossible to express toward older authority figures, particularly fathers.

A close examination of the many instances of animal cruelty suggests nine broad motivations for this behavior, many associated with violent family conditions during childhood:

- *The excessive urge to control animals.* Cruelty toward animals sometimes occurred as a way of controlling an animal's behavior. These situations involved not just exacting compliance, but violently eliminating the animal's "undesirable" characteristics. One man reported kicking his dog in the testicles to stop unwanted behavior; another indicated rubbing his animal's anus in turpentine to rid the creature of certain conduct.
- *Retaliating against an animal.* Some respondents reported inflicting extreme punishment on animals as revenge against some wrong on the part of these creatures. One person recounted shooting and killing a dog that attempted to mate with his animal; another reported burning a cat that had scratched him. Although there was some provocation, the delight taken in the punishment and the intensity of the vengeance suggest cruel and excessive behavior.
- *Satisfying a prejudice against a species or breed.* Cruelty toward animals sometimes reflected extreme bias or hate of particular species

or breeds of animal. The men in our study often directed this behavior at cats. Acts of cruelty toward wild animals often focused on snakes and rodents such as rats and mice.

- *Expressing aggression through an animal.* Sometimes the cruelty occurred as a way of inflicting violence against people. Some of these men seriously abused animals as a way of instilling in them violent tendencies to be used against other people. One man fed his dog gunpowder to make it more violent; another abused his animal to encourage it to attack others at the least provocation.
- *Enhancing one's aggressiveness.* Some of the men routinely abused animals as a way of improving their aggressive skills. One reported maiming creatures for target practice; others harmed animals as a way of impressing people with their capacity for violence. One man repeatedly killed animals as a means of frightening others.
- *Shocking people for amusement.* Animal cruelty sometimes occurred to amuse people. One man reported stuffing cats in a pillowcase, soaking them with lighter fluid, setting them on fire, and then releasing the animals in a bar for a joke. Another man recounted entertaining others by exploding and cutting the legs off frogs.
- *Retaliating against another person.* Extreme cases of animal abuse sometimes occurred as a way of exacting revenge against other people. One man reported retaliating against his neighbor by maiming and killing her pets; another castrated a raccoon and hung its testicles on the door of a woman who had rejected him.
- *Displaced hostility from a person to an animal.* Some of these men inflicted cruelty on animals as a way of displacing aggression from people to relatively defenseless creatures. Many of the violent criminals had extremely chaotic and violent childhoods and were routinely abused by family members. Often these men reported excessive hatred of fathers too powerful to retaliate against. Cruelty toward animals offered these men an opportunity to vent their rage, frustration, and desire for revenge.
- *Nonspecific sadism.* A small minority revealed no obvious motivation for inflicting extreme injury on animals other than the pleasure taken from causing pain. For these people, sadistic gratification appeared to derive from the exercise of power over another

> creature's life. The act of killing seemed to become a satisfying end in itself: pleasure accrued from inflicting pain or extinguishing a life independent of any particular feeling for the animal or other people. In some cases, the satisfaction appeared to be associated with compensating for feelings of personal weakness.

Our study clearly reveals an association between childhood cruelty toward animals and aggressive and antisocial behavior among violent criminals in adulthood. The results suggest that a debased relationship with living creatures can foster pervasive alienation and disaffection. Conversely, cultivating a more benign relationship with animals might promote a more satisfying experience of self. If excessive violence toward animals can become generalized to include other humans, perhaps a more nurturing ethic of compassion and kindness toward nonhuman life can encourage a more gentle attitude toward other people and society.

Conclusion

This chapter has explored various ways people interact with animals and nature and how these varied experiences can influence people's attitudes toward the living world. We have considered a range of consumptive and nonconsumptive uses of wildlife, as well as direct and indirect contacts with nature and animals. Although these relationships may be motivated for many reasons, the quality of the experience has the power to shape not only our perceptions, concerns, and respect for nonhuman life but also, perhaps, how we view ourselves and society.

CHAPTER 5

Species

THIS CHAPTER explores how people's perceptions of diverse species influence their values of nature and biological diversity. People respond in varying and sometimes predictable ways to different categories of life from wolverines to weevils. After examining some of the reasons for these differences, we will consider people's attitudes toward wolves, whales, and invertebrates, insects in particular.

Most attitudes toward animals are a consequence of four major factors (Figure 14). First, the nine basic values of nature and wildlife represent a major influence which, as we have seen, reflects demography, experience, and activity. In other words, people are disposed to view certain creatures in certain ways as a consequence of well-established values of nature. Second, attitudes toward species are also shaped by a creature's particular physical and behavioral characteristics: its size, aesthetic appeal, intelligence, sentience, similarity to humans, cultural and historical familiarity, body shape, and means of locomotion. The third influence reflects people's knowledge of certain creatures. Knowledge

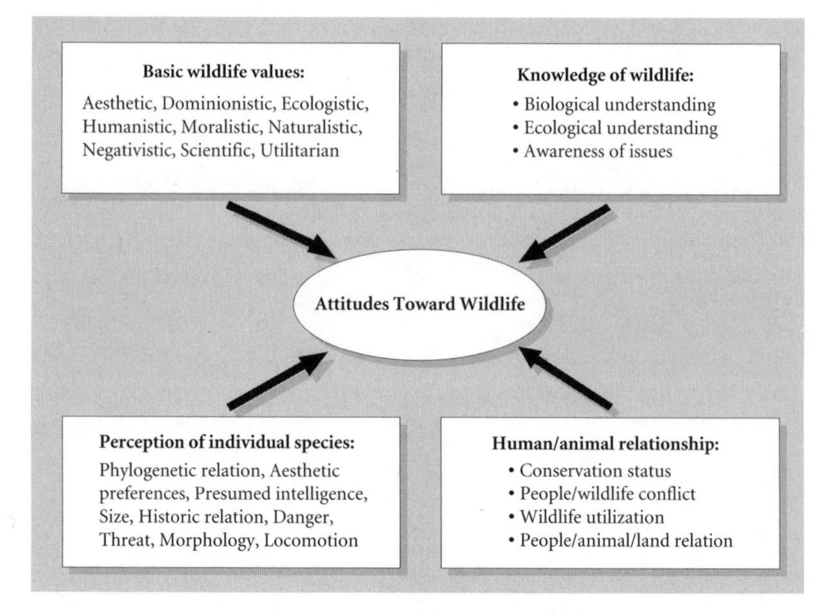

Figure 14. Factors Shaping Attitudes Toward Wildlife

can range from simple factual understanding, to more conceptual and ecological comprehension, to awareness of a species' population and conservation status. The fourth element shaping attitudes toward animals derives from human/animal relationships. These interactions include economic and recreational uses, whether a species occurs on public or private lands, historical treatment of a species, as well as prevailing management practices toward certain creatures.

All four factors shape people's attitudes toward particular species. The combined interaction of these elements may explain why Americans, for example, view bald eagles so differently from harpy eagles, or rattlesnakes from cobras, or caimans from crocodiles. They may also explain why most people view snakes at considerable variance from how they perceive songbirds or grazing ungulates. One's perception of a species, then, reflects a range of influences including knowledge, values, experience, culture, history, and biology. And in each case, the consequent attitudes have much to do with the treatment and destiny of particular creatures.

We partly examined American preferences for diverse animals by querying a nationwide sample about their feelings toward thirty-three well-known species.[1] This examination revealed that the two most preferred animals were domestic species: the dog and the horse. The cat, by contrast, ranked much lower in popularity, reflecting widely divergent views of this creature. Despite the dog's lofty position as America's most favored species, people expressed considerable ambivalence toward its wild cousins, the wolf and coyote, who were represented in the bottom half of the ranked animals. Although little genetic variation may separate dogs from wolves, these two creatures occupy very different psychological niches in most people's minds. Large wild predators—wolves, coyotes, bears, and others—were generally viewed with considerable ambivalence. This wide diversity in views is examined in more detail later in the chapter when we consider past and present attitudes toward the wolf.

The two most preferred wild animals were birds: the robin and swan. Birds constituted the most liked of all taxa, mammals included. The most popular fish species were the trout and salmon, indicative of the cultural and historical importance of these creatures. Despite the general significance of biological similarity to humans in shaping people's attitudes toward animals, all game species were viewed positively whether mammals, birds, or fish. The favorite invertebrates were the butterfly and ladybug, suggesting the role of aesthetics in shaping attitudes toward animals. The best-liked nonindigenous species were the elephant and lion, indicative of the significance of size in people's preference for animals. The most appreciated herpetological species were the turtle and frog.

The most disliked animals were two invertebrates: the cockroach and mosquito. In fact, invertebrates constituted the least appreciated of all taxonomic groups. With the exception of creatures like butterflies and bees that possess redeeming aesthetic or instrumental virtues, most insects aroused negative sentiments. The role of these and other factors in attitudes toward invertebrates, particularly insects, is examined later in the chapter. The rat was the most reviled mammal. Among reptiles, birds, and fish this dubious distinction was represented, respectively, by the rattlesnake, vulture, and shark. Hostile views were directed at bats, as well, possibly indicative of the role that superstition, myth, and nocturnal behavior play in our perceptions of animals.

The species were also grouped according to varying taxonomic, physical, and human relational characteristics. The most preferred groups were domesticated, aesthetically appealing, and game animals. The least preferred were the biting and stinging invertebrates, aesthetically unattractive species, and animals associated with human injury, disease, and property damage. Birds and mammals comprised the favorite vertebrate classes. Invertebrates tended to be regarded as alien and unappealing.

These results reflect the various physical, behavioral, and human relational characteristics cited earlier as influencing attitudes toward animals. Cultural and historical familiarity, as well as the economic and recreational significance of a species, clearly play an important role in people's perceptions. The biological similarity of the species to human beings represents a significant influence, too, including assumptions about an animal's intelligence, sentience, and capacity for pain and suffering. The relationship of the species to human society looms important—especially the animal's status as a pet, game species, livestock, pest, or native versus foreign animal. Another major factor is the creature's aesthetic appeal, greatly influenced by such considerations as color, shape, movement, and visibility. Other key characteristics include a species' skin texture, way of moving, and size. A hypothetical composite gleaned from these various characteristics would suggest that people generally prefer large attractive animals with an erect bearing, animals that walk, run, or fly rather than crawl, slither, or live underground. A good candidate for the average human nightmare might be a creature that is small, ugly, predatory, likely to inflict injury and property damage, lacking in intelligence and feeling, and a denizen of dark, damp places, inclined to crawl and slither about.

Japanese and Germans revealed very similar preferences in analogous studies in those countries. Although the Japanese evinced more favorable views of invertebrates, whereas the two Western nations generally expressed stronger preference for large mammals, the general similarity of species preference in all three nations suggests the biological origin of human attitudes, at least toward well-known creatures. Perhaps people's animal preferences reflect physical, intellectual, and emotional adaptations emerging over the course of human evolution.

Despite this possibility, culture and history clearly influence attitudes toward certain creatures. This chapter explores in detail the role of such factors in past, present, and changing perceptions of the natural world. Three particularly revealing animal groups are used to illustrate this discussion: wolves, whales, and invertebrates, mainly insects. Wolves on land, and whales in the sea, evoke strong reactions among Americans and, more generally, Westerners. These attitudes have shifted dramatically in recent times, though, rendering wolves and whales especially sensitive barometers of changing Western views of the natural world. Invertebrates have largely inspired hostile attitudes. Exploring these negative views and occasionally more positive sentiments may tell us something about how humans perceive nature and living diversity.

The Wolf

The wolf has inspired more passionate, ambivalent, and contrasting attitudes among Americans than perhaps any other creature.[2] From the very beginning of the European settlement of North America, the wolf provoked powerful sentiments. Indeed, the wolf was the dubious recipient of the American colonies' first official wildlife act: a one cent bounty levied by the pilgrims of Plymouth County in 1630. Most Native Americans also viewed the wolf as a creature of great power, but more typically as a source of inspiration rather than an evil presence or vicious competitor.

The American pioneer inherited hostile attitudes toward wolves from Europe, where the animal had been persecuted for centuries. The wolf was especially despised by the European settlers because it suggested a threat to personal safety and an impediment to progress and civilization. The symbolic connection of the wolf and the Native American only reinforced in the settler's mind an inclination to conquer the hated wilds and exercise mastery over the continent's large predators and its indigenous peoples. Wilderness represented an obstacle to subdue, render productive, or eliminate altogether. Subjugation of wolves and wild places merged through this ethical prism as an expression of duty, reflecting the settler's will to dominate the land and transform it into something useful.

Attesting to the prevailing wisdom of the time, John Adams remarked in 1756: "The whole continent was one continuing dismal wilderness, the haunt of wolves and bears and more savage men. . . . Now the forests are removed, the land covered with fields of corn, orchards bending with fruit and the magnificent habitations of rational and civilized people."[3]

The wolf and other hateful creatures like the rattlesnake, rat, and coyote tended to be viewed from the perch of this morality play as intrinsically evil. Destroyers and vermin by nature, these creatures were viewed as having been placed in the settlers' path as a test of their faith and fortitude. Killing wolves signified one's belief in community and godly obligation as much as a response to presumably practical threats to livestock and person. Total eradication seemed both ethically justified and pragmatically responsible according to this worldview. As Stanley Young recollects: "There was a sort of unwritten law of the range that no cow man would knowingly pass by a carcass of any kind without inserting in it a goodly dose of strychnine sulfate, in the hope of killing one more wolf." [4]

The fever pitch of wolf killing reached a peak during the settlement of the western prairie, a time when wolves initially flourished following the slaughter of the bison, antelope, and other large ungulates, and then as domestic stock began to replace the wild herds. Loathing for the wolf eventually achieved the scope of a national extermination campaign bent on ridding American civilization of this evil presence once and for all. The wolf became both progressive symbol and scapegoat. As Barry Lopez observes: "The wolf was not the cattlemen's only problem—there was weather, disease, rustling, fluctuating beef prices, hazards of trail drives. . . . [But] the wolf . . . became an 'object of pathological hatred.' . . . Men in a speculative business like cattle ranching singled out the wolf as a kind of scapegoat for their financial losses. . . . It was against a backdrop of . . . taming wilderness, the law of vengeance, protection of property, an inalienable right to decide the fate of all animals, . . . and the . . . conception of man as protector of defenseless creatures—that the wolf became the enemy."[5]

Even after the wolf had been extirpated in one state after another, the animal's mythic power remained. In 1896, a cattleman could still proclaim: "The number of wolves has become so considerable that all means of extermination . . . have only succeeded in keeping them at a standstill."[6] Less than a quarter of a century later, the wolf problem had been solved, the animal for all practical purposes eliminated from the forty-eight contiguous states. Loathing for the creature even surfaced among such early champions of wildlife protection as President Theodore Roosevelt, who described the wolf as "the beast of waste and desolation," or William Hornaday, president of the New York Zoological Society (and, ironically, a strident opponent of the leg-hold trap), who remarked: "Of all the wild creatures of North America, none are more despicable than wolves. There is no depth of meanness, treachery or cruelty to which they do not cheerfully descend."[7] The wolf's destruction became official government policy, a seemingly unavoidable price of modern civilization. As William Goldman, director of the Bureau of Biological Survey, suggested: "Large predatory mammals, destructive of livestock and game, no longer have a place in our advancing civilization."[8]

As the extent and viciousness of the killing often reached irrational proportions, one suspects the wolf may have performed roles beyond the merely utilitarian. Destroying the wolf may have also reflected the urge to rid the world of an unwanted and feared element in nature, perhaps even the settler's atavistic potential to succumb to the allure of wildness and the absence of civilizing control. Further, wolf hatred suggests an ignorance of human dependence on nature and living diversity for a variety of ecological, emotional, intellectual, and spiritual purposes. Lopez speculates on these varying sources of wolf hatred: "The motive for wiping out wolves proceeded from misunderstanding, from illusions of what constituted sport, from strident attachment to private property, from ignorance and irrational hatred. But the scope, the casual irresponsibility, and the cruelty of wolf killing is something else. I do not think it comes from some base, atavistic urge, though that may be a part of it. I think it is that we simply do not understand our place in the universe and have not the courage to admit it."[9]

American attitudes toward wolves and wilderness have changed considerably during the course of the twentieth century. Perhaps the increasing rarity of each has expanded the value of both, although it may also reflect a maturing of thought and an expansion of knowledge. A dramatic shift in the attitude of Aldo Leopold toward wolves anticipated a general change in society's perceptions of this animal.[10] At the outset of his career, Leopold like most of his contemporaries viewed wolves as an expendable element in nature. His initial assignment in America's first wilderness area, the Gila National Forest of New Mexico, included extirpating wolves in the hope this would result in greater numbers of deer and other ungulates, thereby creating the conditions for better human hunting. The experience of killing wolves, however, changed Leopold's views of both this animal and wilderness, becoming an epiphany in his evolving notions of ecology and ethical obligations to the natural world. Leopold conveys aspects of his shift in attitude and understanding of wolves and wilderness in this passage:

> In those days we had never heard of passing up a chance to kill a wolf. . . . I was young then. . . . I thought because fewer wolves meant more deer, that no wolves would mean hunters' paradise. . . . Since then I have lived to see state after state extirpate its wolves. . . . I have seen every edible bush and seedling browsed, first to anemic desuetude, and then to death. . . . I now suspect that just as a deer herd lives in mortal fear of its wolves, so does a mountain live in mortal fear of its deer. . . . Perhaps this is the hidden meaning in the howl of the wolf, long known among mountains, but seldom perceived among men.[11]

The wolf also emerged for others as a potent symbol of the enduring wisdom of wilderness, and a lament over America's legacy of despoliation and destruction of the continent's wildlife. Especially in the years following World War II, the wolf became one of the nation's most powerful icons of wildlife preservation and an atonement for the country's past persecution of its natural diversity. The wolf was among the first creatures to be listed as in peril of extinction following the passage of the Endangered Species Act. An extraordinary number of books, films, and television specials celebrated the nobility of this animal and helped to transform wolves from an evil and loathsome presence into a worthy, intelligent, and admirable creature.[12]

Contemporary attitudes, however, remain at the cusp of a transition in the perception of wolves, wildlife, and wilderness. Many now view wolves as innocent victims of a society that lost its bearings and place in nature; others continue to hold antagonistic values toward this animal and other large predators.[13] Deeply embedded skepticism toward wolves, even hostility, especially prevails among those living in proximity to existing or proposed wolf populations including Alaska, Minnesota, Michigan, Montana, Wyoming, and New Mexico. Local opponents of wolf reintroduction to Yellowstone National Park have suggested: "The wolf is like a cockroach and will creep outside of Yellowstone and devour wildlife." Or: "Only a brain-dead son-of-a-bitch would favor reintroduction of wolves. It's like, it's like, inviting the AIDS virus." Or: "Wolves don't feed and water the livestock and they don't help raise food for people to eat, so what good are they?"[14] Advocates of Yellowstone wolf restoration, on the other hand, have countered: "Only a fool would not agree to the placement of this beautiful and essential animal." Or: "Wolves do not kill people. Fatty beef does."

We conducted more systematic assessment of contemporary attitudes toward wolves in studies in Minnesota and Michigan. Our nationwide study had previously revealed that Americans divided almost evenly in their negative and positive views of wolves. Dislike of the wolf was especially prominent among livestock producers, elderly persons, rural residents, and the least educated. Far more positive sentiments toward this animal emerged among young adults, the college-educated, city residents, and members of environmental organizations.

The Minnesota and Michigan studies provided detailed information on perceptions of wolves, their conservation, and management. Minnesota is the only one of the forty-eight contiguous states with a wolf population that has never been extirpated: some 1,200 wolves remain. Michigan, however, has only three dozen or so wolves who naturally recolonized the state's Upper Peninsula after being absent for many decades. A deliberate attempt in the 1970s to reintroduce wolves to the Upper Peninsula failed largely because of human-inflicted mortality. A study conducted after the unsuccessful reintroduction revealed that deeply ingrained antipredator and antigovernment attitudes contributed significantly to the effort's failure.

We measured attitudes toward the wolf according to the typology of values toward wildlife. Our scales consisted of statistically clustered questions derived from a 30- to 40-minute survey of 621 Minnesotans and 639 Michigan residents. Each study included random samples of the general public, farmers, hunters, and trappers. Given the similarity of results in both states, only the Minnesota findings are highlighted here.

As Figure 15 reveals, widely divergent attitudes toward wolves appeared among Minnesota's urban residents, farmers, deer hunters, trappers, and northern rural people living in proximity to wolves. Farmers expressed especially antagonistic attitudes toward wolves, viewing this animal as possessing little ecological, recreational, or ethical value. Moreover, these hostile sentiments differed little among Minnesota farmers whether they lived near existing wolf populations or areas that had not seen a wolf in half a century or more. These and other study results consistently reveal deeply entrenched bias among agriculturists, particularly livestock producers, toward wolves and other large predators, often independent of personal experience.

Although residents of northern Minnesota rarely expressed as much antagonism toward wolves, most voiced more negative and less appreciative views than found among urban residents. Most city and suburban dwellers, particularly younger, better-educated, and higher-income persons, viewed the wolf as possessing considerable ecological and recreational value and strongly endorsed this animal's right to exist and its protection and restoration to the wild. Rural Minnesotans expressed appreciation of the wolf, but they also viewed this animal as a competitor whose restoration often represented a threat to their economic livelihood and way of life.

Hunters and trappers revealed somewhat unusual views of wolves. On the one hand, they endorsed the right of humans to exercise mastery over this animal, to control wolves that conflicted with human interests, and to kill wolves for practical purposes when wolf populations supported some degree of consumptive use. On the other hand, they expressed strong affection for this animal, emphasizing it's ecological importance, and supported efforts to restore the wolf to its rightful place in the northern forest.

The Michigan study revealed a gap between attitudes and knowledge. Despite fundamental differences in attitudes toward wolves among people who supported and opposed wolf restoration, no significant variations emerged among these antagonists in respective knowledge of this

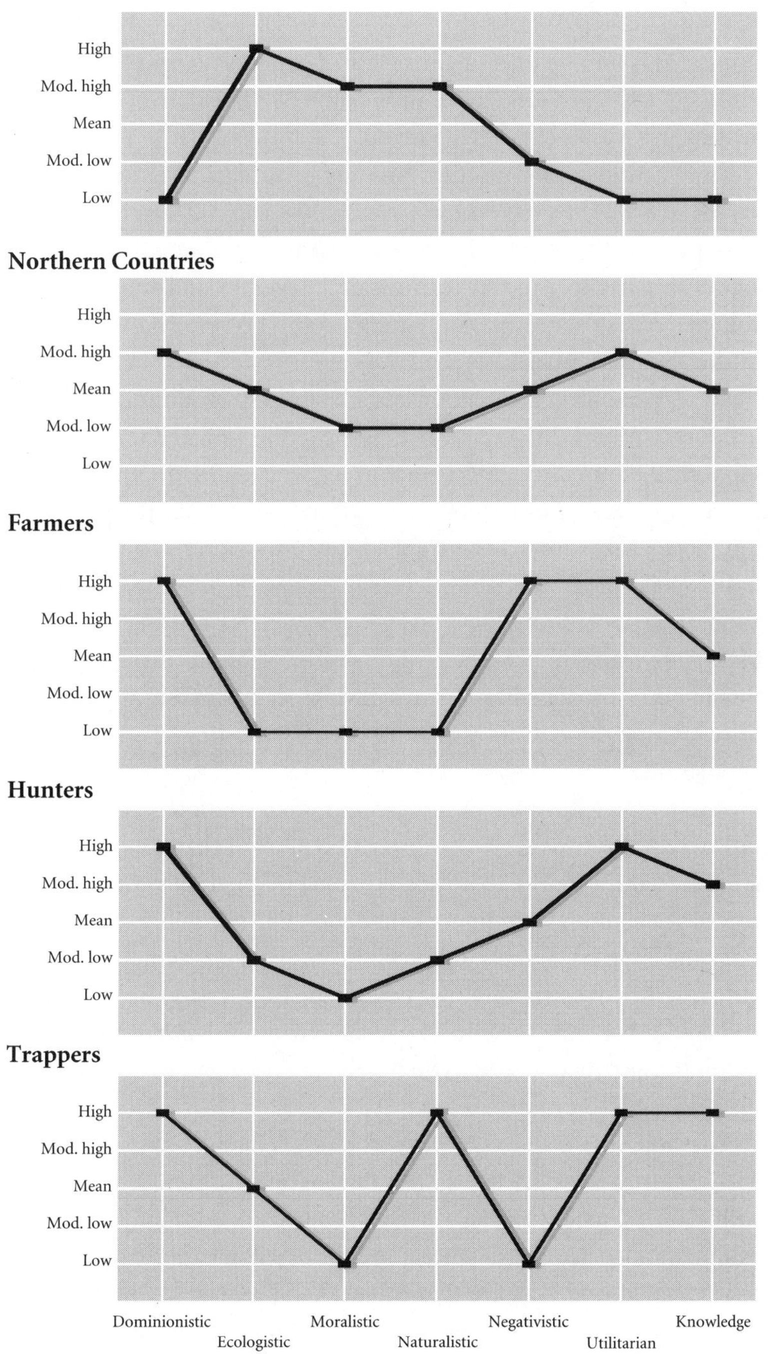

Figure 15. Attitudes Toward Wolves in Minnesota

animal. This finding once again indicated people's tendency to use additional knowledge to rationalize and reinforce rather than change strongly held attitudes toward the living world.

Thus considerable variation prevails in American attitudes toward wolves and their conservation. Farmers, the elderly, the rural, and the less educated continue to view this animal as threatening, as lacking ecological importance, and as intrinsically unworthy and sometimes evil. Younger, college-educated, and urban people, however, have come to regard wolves and other large predators as especially noble and admirable creatures possessing great moral and naturalistic significance. Most contemporary Americans, even those less favorably inclined toward wolves, support this animal's rightful place in the wild, and most view this creature as a reflection of nature's wonder and beauty. Yet the majority of Americans also support the right of farmers to protect their livestock from wolf depredation, although they believe such controls should be humane and focus only on wolves responsible for causing the damage. Most Americans support the goal of wolf restoration, but few think this should occur by excluding various economically significant activities or by restricting the right of people to live near existing or proposed wolf populations.

This country's history of intense and often irrational persecution of wolves has made this animal a powerful symbol of changing attitudes toward the natural world. Many today regard wild nature as the expression of a delicate and nearly divine handiwork, and the living species, especially the large carnivores and predators, as vital components of a complex, quasi-living structure. For others, the wolf, grizzly bear, mountain lion, and other large predators remain a vivid reminder of the necessity to combat and repress wild nature in the never ending struggle to render the land safe and productive.

The wolf once inspired nearly pathological hatred—indeed, its elimination was considered the destruction of an evil presence at variance with civilized society. Today the wolf has become for many a symbol of wildness and an icon in an emerging ethic of wildlife conservation. These conflicting attitudes assure that the wolf will remain a powerful litmus test of shifting American attitudes toward nature and living diversity. The wolf, like so many other charismatic species, continues to represent significant dimensions of meaning, quality, and symbol to human society. By preserving this species, these values are protected as well, for people depend on a broad array of relations with the living world to experience lives filled with meaning and purpose. Wolves and other large predators will hopefully remain irreplaceable contributors to human language, intellect, story, myth, and connection with the living world.

The Whale

Barry Lopez, searching for a metaphor to describe the somber history of the wolf in America, aptly compares it to the whale when he suggests that men went after wolves like Ahab after the whale.[15] Wolves and great whales share the dubious distinction of being among the most persecuted of animals during the eighteenth, nineteenth, and early twentieth centuries, each brought to the brink of extinction. Whales, like wolves, also became the focus of a profound attitudinal transformation—a symbol emerging during the latter twentieth century as a powerful rallying cry for a developing ethic of wildlife protection and restoration.[16] The whale in the sea and the wolf on land have become commanding tokens of atonement for the excesses of human greed and arrogance toward the living world.

The treatment of whales and wolves differed significantly, however, in the motive for each animal's persecution. The killing of whales, unlike wolves, proceeded more from avarice than hatred.[17] The wolf had largely been a reviled creature, a crepuscular scavenger after the dead, an animal that presumably killed for sheer pleasure and blood lust. Its utter annihilation struck many as both economically warranted and morally justified. Whales, by contrast, rarely achieved such infamy or presumption of wickedness, although many perceived the great leviathan as a monster in the sea. The primary motive for decimating whales was venal. Still, a secondary drive may have been the desire to conquer the most prodigious of all creatures in the most alien and challenging of environments and, through courage and determination, render both the creature and its habitat the product of practical commerce.

The whale was nonetheless viewed by most as a noble adversary, nearly the equivalent of man. The whale inspired sentiments of admiration, even awe and respect, quite unlike the intense fear and loathing at the core of the wolf's destruction. For both creatures, however, the end result was nearly the same: their near elimination and the submission of each to human will and desire. The whale in the sea, like the wolf on land, constituted not only a symbol of wildness but also a fulcrum for projecting attitudes of conquest and utilitarianism and, eventually, more contemporary perceptions of preservation and protection. To understand attitudes toward whales we must first examine the forces that led to their overexploitation and decimation. Only then can we begin to understand why emerging attitudes of whale appreciation and protection still coexist along with views favoring whale exploitation and other forms of marine mammal utilization.

Whales and, more generally, marine mammals reflect the devastating effects of human avarice when fueled by ignorance and arrogance. Kenneth Norris indicates the extent of marine mammal destruction when he notes: "No other group of large animals has had so many of its members driven to the brink of extinction."[18] David Ehrenfeld observes that of the three primary orders of marine mammals, eight species of whales and dolphins, ten species of seals and walrus, and five species of manatees and dugongs are extinct or currently threatened with extinction. Of seventy-one species of marine mammals in U.S. waters, seventeen are currently listed as endangered and two as threatened with extinction. The great whales, particular victims of overexploitation, are nearly all endangered today despite years of protection. Only the Eastern Pacific gray whale has recovered, perhaps because of its newly elevated status as the major focus of a new human use: the recreational observation of whales in the wild.

The historic slaughter of the great whales has sometimes been called the greatest fishery the world has ever known. Three distinct periods marked the whale's exploitation. The initial whaling era, which occurred prior to the eighteenth century, consisted mainly of coastal whaling. This period focused on "right" whales—slow coastal creatures who floated when struck and thus constituted the "right" whales to kill. The second era centered on the New England whaling industry, a time when the great sailing ships roamed the seven seas in search of what, along with another wild animal, the beaver, would produce some of America's first millionaires. The third whaling era, the modern period, marked the triumph of humans over whales, a time when modern technology—the explosive harpoon, pneumatic lance, fast catcher boats, factory ships, and other technical innovations—permitted the exploitation of any whale no matter how fast or pelagic and all to be driven to near extinction. The killing of whales peaked during the early 1930s with over 60,000 animals killed in some years. Measured in overall weight, however, the height of the slaughter occurred as recently as 1960 with the harvest of nearly 4 billion tons of whale meat or about 15 percent of the world's marine production that year.

Until recently, the great whales were largely managed as creatures whose value consisted only of commodities and economic returns. This overwhelmingly utilitarian orientation was reflected in a killing standard, the "blue whale unit," maintained until 1972, which measured the worth of any whale species by its weight in relation to blue whales as if one species pretty much equaled another. Reflecting this economic orientation, protection became the preferred option only after the great whales had become, for all practical purposes, commercially extinct. As Peter Matthiessen once sourly remarked: "The removal of financial incentives . . . makes conservationists of one and all."[19]

Apart from technology, a number of other factors contributed to the decimation of the great whales and other marine mammals. A frequent problem in effectively conserving these creatures is the difficulty of controlling the behavior of individual fishing boats and nations in the world's oceans: restricting access, effecting control, allocating harvesting rights. Lacking jurisdictional authority, the protection of whale stocks has remained problematic, as has the enforcing of international agreements. Most whales behave according to Garrett Hardin's "tragedy of the commons,"[20] that is, the property of no one and, therefore, everyone. This situation—many independent whaling boats representing the interests of diverse nations, each having access to pelagic whale stocks and possessing efficient killing technology—produced few incentives for restricting the harvesting of whales or allowing the effective monitoring and accurate reporting of actual levels of exploitation.

The great whale—the most prodigious creature the world has ever known in the vastness of the world's oceans—conveyed the impression of being an inexhaustible resource. This perception was reinforced by the presumption that one whale, at least from an economic perspective, was very much like another. The twin fallacies of inexhaustibility and substitutability represented a poisonous combination encouraging whalers to eliminate one whale stock after another.

Ignorance and arrogance contributed to the great whales' demise as well, although these animals were among the first species to receive some form of international protection—whaling conventions signed by the whaling nations first in the 1920s and then following World War II. Until a moratorium on whaling was instituted in the 1980s, the only periods of significant whale recovery ironically occurred during the two world wars, times when people appeared to be too busy killing one another to slaughter whales. The commercial objectives of the whaling conventions blinded almost all to the unknowns of whale biology or the other values represented by a mammalian order with a longer and more successful evolutionary tenure than its human tormentors. Greed and avarice, the burden of major investments in expensive technology, the whale's slow reproductive rate, and a host of other factors eroded and subverted the principle of sustainable yield. In fact, there was a far more compelling logic: overexploit the animal and reinvest the surplus profits in other areas of commercial activity.

The great whales' demise was hastened, too, by unsympathetic attitudes. Whales were viewed mainly from the perspective of the great fishery, a place where men dared and defied the alien seas in the search for great riches. The whale's conquest reflected the human capacity to take on the most challenging of creatures in the most defiant of environments, bending each to the will of people and the presumed needs of society. Whales were largely the victims of strongly held utilitarian, dominionistic, and negativistic values. But a profound change in North American and European attitudes emerged during the second half of the twentieth century. For many, whales became the focus of ecologistic, naturalistic, humanistic, and moralistic values. The whale became a powerful metaphor for an evolving ethic of wildlife protection and wilderness restoration.

Many factors contributed to this change. A substantial increase in knowledge and understanding of whales certainly encouraged a shift in attitudes.[21] The years following World War II witnessed a remarkable increase in scientific study of whales spurred by the development of innovative technologies for studying the ocean and the requirements of unprecedented legislation such as the Marine Mammal Protection Act and the Endangered Species Act. New knowledge conveyed a new impression of whales as animals of great intelligence, complex social lives, highly sophisticated powers of communication, and other characteristics suggesting fundamental similarities between cetaceans and humans. This knowledge evoked feelings of admiration, even awe, for these animals at the apex of their food chain who had so successfully mastered the challenge of survival. People learned about the "singing" of humpback whales, the "talking" of dolphins, the "kindness" of killer whales, and many came to regard these creatures through a far more appreciative lens. Many even viewed whales as species akin to humans, certainly not brute, inferior beings whose primary value derived from their material utility to people. The highly endangered status of most great whales—and mounting concern for the world's declining biological heritage—also made whales a powerful symbol for protecting imperiled wildlife.

Human uses of whales also changed dramatically from whaling to whalewatching, and a large and economically powerful constituency emerged far more interested in whales alive than dead.[22] Whalewatching initially developed during the 1950s along the California coast, focusing on the migration of the gray whales from the northern arctic seas to the breeding and calving lagoons of the Baja peninsula. The popularity of whalewatching soon spread to other regions of the United States, especially New England, Alaska, and Hawaii. By 1981, whalewatching grossed some $14 million annually; by 1988, whalewatching had become a worldwide industry with revenues approaching $50 million. A recent survey estimated that $320 million was expended on whalewatching in 1992 in more than thirty countries. In a stunningly short time, whales had been transformed from a consumptive to nonconsumptive resource.

The international management of whales also shifted during this period.[23] A global moratorium halted most commercial hunting of whales in the 1980s, although Norway, Japan, Iceland, and others continued to object to the ban and practiced limited "scientific" whaling.

Major legislative change was enacted, as well, particularly in the United States. Whales had historically been managed as a commodity resource using the principles of livestock and fisheries exploitation. The passage of the U.S. Marine Mammal Protection Act in 1972, however, and the subsequent establishment of a 200-mile zone of economic control along America's coasts, dramatically shifted the basis for managing and protecting whales. The act's bold provisions included:

- A moratorium on the taking of marine mammals in U.S. waters with rare exceptions (by Native Americans, for example, or for scientific or educational purposes)
- A broad definition of "taking" including harassment and habitat loss, as well as direct capture, injury, or mortality
- An emphasis on protecting individual whale populations as well as entire species
- Creating an "optimum sustainable population" standard to replace the traditional fisheries objective of "maximum sustainable yield"
- Preempting state authority to manage marine mammals in favor of federal jurisdiction and control
- Imposing sanctions against other nations who violated U.S. marine mammal protection laws
- Requiring that marine mammals be captured in ways which minimized suffering and pain

For many Americans, therefore, the great whales had been conceptually transformed from objects of commerce to subjects of recreational interest, ecologistic admiration, protective concern, and ethical respect. As in the case of the wolf, however, considerable variation in attitude remains. In fact, the historic transition is incomplete, for significant elements of the population are still inclined to view whales from the traditional utilitarian and dominionistic perspectives. Whales continue to be caught in the transition between a past biased by an exploitative mentality and a future characterized by more compassionate attitudes toward living diversity.

Variations in perceptions of whales and other marine mammals are reflected in a study we conducted in Canada.[24] This study included samples of commercial sealers, fishers, and the Canadian general public

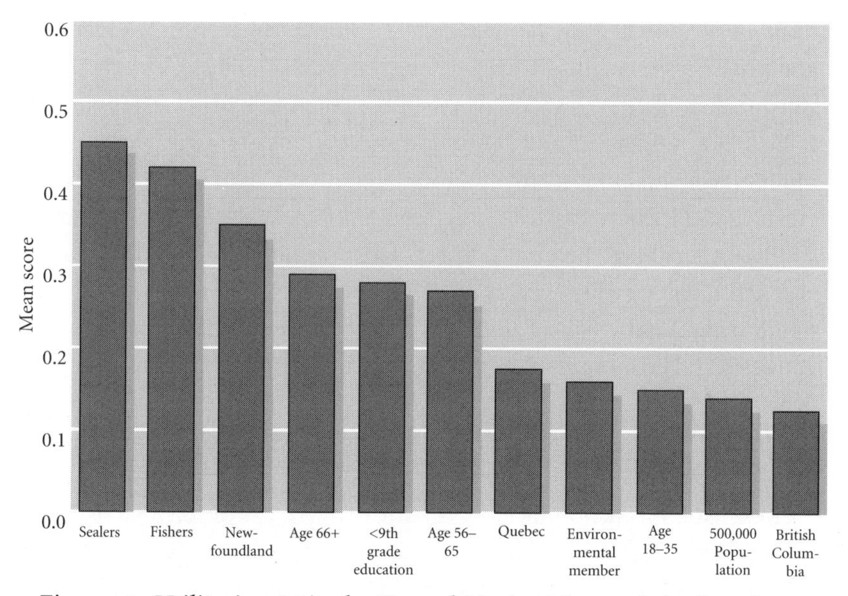

Figure 16. Utilitarian Attitudes Toward Marine Mammals in Canada

Figure 17. Moralistic Attitudes Toward Marine Mammals in Canada

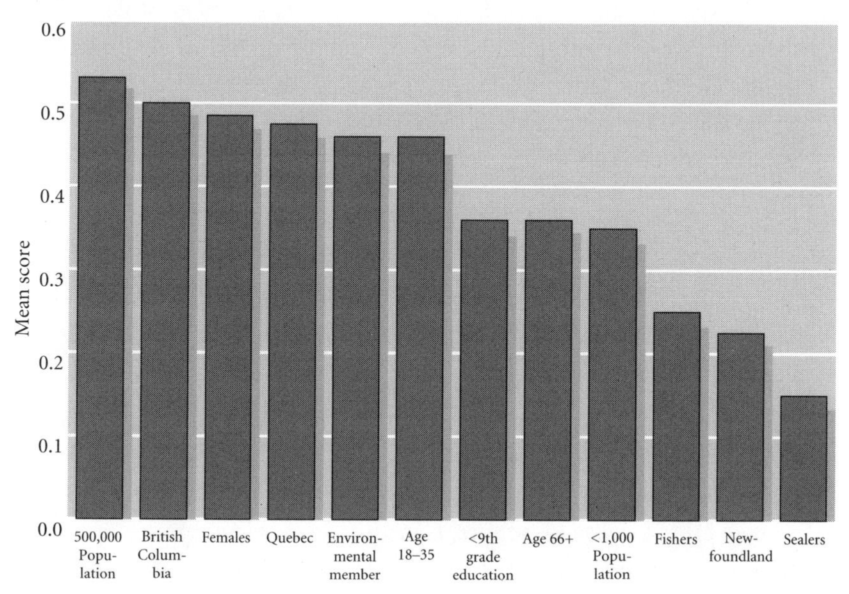

distinguished by age, gender, education, income, urban/rural residence, and province. As Figures 16 and 17 suggest, rural residents, elderly persons, sealers, fishers, and residents of the Atlantic maritime provinces hold strong utilitarian and dominionistic attitudes toward whales and other marine mammals, largely endorsing people's right to harvest these creatures and assert mastery over them. Urban, college-educated, and younger Canadians, however, express far more moralistic and humanistic attitudes, viewing whales as subjects of affection, recreational interest, and ethical concern, even when these values conflict with human activities and economic interests. These demographic groups also differ in their views of whale conservation, ocean development, and commercial fishing. Only small minorities of the college-educated, younger, and urban Canadians, in contrast to large majorities of sealers and fishers, for example, support economically exploiting whales or commercial fishing if it harms marine mammals.

A six-nation study of attitudes toward whales and whaling, conducted in 1993, provides additional comparisons among the United States, England, Germany, Norway, Japan, and Australia. In the Canadian study cited earlier, nearly two-thirds of the public objected to whaling under any circumstance in contrast to one-quarter of commercial sealers and fishers. In the six-nation study, only one-third or less of Australians, British, Germans, and Americans endorsed the notion of hunting nonendangered whales. A majority of Japanese and Norwegians, like the Canadian sealers and fishers, supported the taking of nonendangered cetaceans.

This study also reveals a profound shift in American attitudes toward whaling in just one decade. In 1980, three-quarters of the American public endorsed the hunting of nonendangered whales if a useful product was obtained. But in the six-nation study, done in 1993, less than one-third of Americans approved of whaling under this circumstance, as well as even smaller percentages of Australians, English, and Germans (Figure 18). Two-thirds to three-quarters of Japanese and Norwegians, however, approved of killing plentiful whales for practical purposes. The Japanese and Norwegian publics, Canadian sealers, commercial fishers, and Newfoundland residents, therefore, endorse whaling if properly reg-

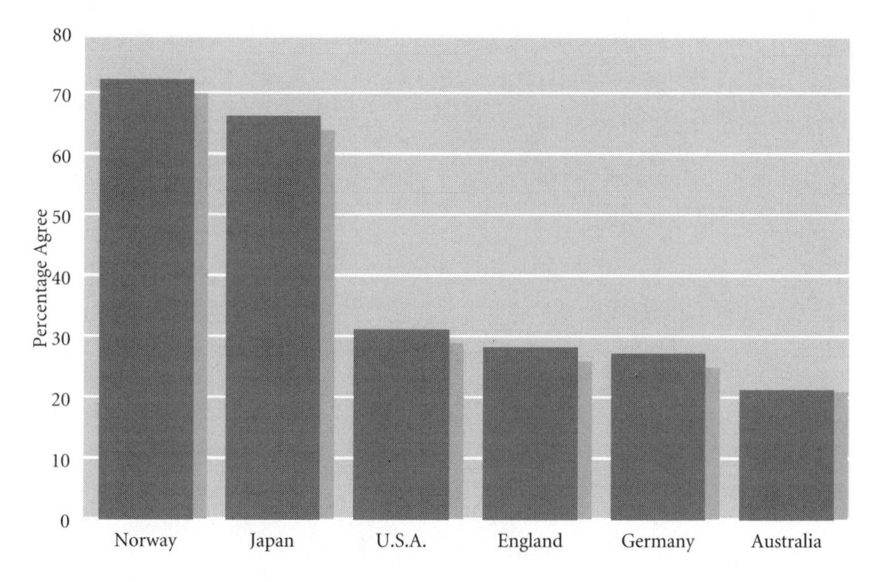

Figure 18. Response to, "If Whales Were Plentiful Again, Harvesting for Useful Products Would Be Acceptable."

ulated, if the species is not endangered, and if it yields significant economic and cultural benefits. Antiwhaling sentiment, under almost any circumstance except aboriginal whaling, was expressed by the majority of Americans, English, Germans, Australians, and Canadians.

These results reveal pronounced changes in attitudes toward whales. For many these creatures have become compelling symbols of a new and more benign relationship with nature and living diversity. Yet the shift in attitudes is still tentative and in transition. On the one hand, an increasingly educated and urbanized public continues to embrace more positive, protectionist, and nonconsumptive views of whales and other marine mammals. The popular image of whales is likely to become more sympathetic, too, as people learn more about the complex biology and social behavior of these animals, as well as the continuing threats to their existence. On the other hand, support for the commercial exploitation of wild living creatures remains deeply entrenched in the human psyche, especially among certain groups. Moreover, economic and regulatory circumstances could cause a resurgence of support for resuming the harvest of whales and other marine mammals. One conclusion at least is clear: attitudes toward whales will remain a keen barometer of changing conceptions of wildlife and the place of humans in the natural world.

Invertebrates

So far our examination of attitudes toward animals has focused on familiar vertebrates—culturally, economically, and symbolically significant creatures. Most life, however, consists of invertebrates, especially insects. Although estimates vary, 10 million species of plants and animals are thought to exist on earth today. Of this total, more than 90 percent are invertebrates.[25] Even among the roughly 1.7 million species that have been scientifically identified and described, approximately 900,000 are insects and spiders. By comparison, some 4,000 mammal and 9,000 bird species have been identified. The beetles are the most numerous of all the insects, comprising almost 300,000 known species, and the social insects combined (ants, bees, wasps, termites, and others) may account for the majority of life on earth. Most of this insect life occurs in the moist tropics, where three-hundred butterfly species might be encountered in a square kilometer of tropical forest compared to perhaps ten to fifteen in a similarly sized temperate area of North America.

Invertebrates are remarkably important ecologically, but they also bestow considerable practical benefits on people.[26] Some of these ecological and utilitarian values will be reviewed here before we turn to the question of how most people actually view invertebrates. Ecologically, invertebrates perform a striking variety of functions. Invertebrate life remains indispensable in such critical ecological processes as plant pollination and protection, energy and nutrient transfer, species mutualism and competition, ecosystem maintenance and stability, decomposition, predation and parasitism, and sometimes even the provision of entire habitats for many other life forms. Although the exact relationships remain unclear, interactions among invertebrates and between invertebrates and other organisms often appear to help maintain the health and resilience of entire ecosystems. Ecological health can be seriously compromised when large-scale reductions occur in the complexity and diversity of invertebrate life. For example, agricultural and forestry monocultures can result in extensive pest infestations stemming from the elimination of their natural invertebrate predators.

Certain invertebrates, sometimes referred to as "keystone" species, play critical roles in maintaining entire biological communities. For example, species diversity in moist tropical forests is often a consequence of varying interactions and interdependencies among plants and insects, the latter often performing indispensable roles in plant pollination, seed dispersal, and plant protection. Eliminating these insects can result in the decline of entire species whose existence depends on these interconnections.

A constant stock of edible organisms is essential for the existence of any ecosystem. Invertebrates often provide the base of energy and food pyramids because of their extraordinary numbers. For example, the biomass of earthworms and arthropods in a typical northeastern American forest may amount to 1,000 kilograms per hectare, in contrast to 18 kilograms for all other wild terrestrial vertebrates. In many marine environments, the transfer of energy and nutrients from one energy level to another may be performed by invertebrate organisms. A dramatic example is the link between tiny shrimplike krill and the largest creatures on earth, the baleen whales.

Pollination and seed dispersal represent two essential ecological functions often performed by invertebrates, particularly bees, wasps, butterflies, and flies. In North America, some 5,000 bee species pollinate plants, while another 20,000 plant species depend on the cross-pollinating activities of insects. Parasitic and predatory insects also help prevent plant-eating organisms from growing so numerous they become destructive. The elimination of these protective insects in large-scale agriculture and forestry operations has sometimes resulted in major pest infestations and the extensive need to use dangerous pesticides with many known and unknown health hazards.

Invertebrates sometimes provide entire habitats for other organisms. The best-known example is the coral reef—an environment built on the skeletal remains of various invertebrate creatures. Coral reefs have been described as the marine equivalent of the rain forest, accounting for perhaps one-third of all fish species.

A number of human benefits derive from these ecological functions of invertebrates: pest control, maintaining soil quality, waste decomposition, and the provision of various foods, medicines, and other commercial products. Naturally occurring invertebrates help control agricultural pest infestations, resulting in the savings of many billions of dollars annually. Deliberately introduced insects have also been used to control harmful organisms, as well as provide synthesized chemicals to repel pests and attract undesirable organisms into traps.

Many invertebrates help decompose dead organic material. Earthworms and ants, for example, play critical roles in maintaining soil fertility, essential for agriculture and forestry. These and other invertebrate organisms help disintegrate fibrous tissues, mix organic materials with the soil, and increase soil porosity and drainage. Among the most important benefits derived from such invertebrates is the breakdown of the enormous wastes generated by modern society. In the United States, people annually produce an estimated 130 million tons of excreta, while livestock contribute another 12 billion tons of manure. Nearly all of this organic material is decomposed by invertebrates and microbial organisms, leaving one to wonder what humans would be up to their eyeballs in if not for the labors of these little friends.

Some ninety agricultural crops in the United States depend on the pollinating activities of mainly insect species. More directly associated with the human diet, the total "shellfish" catch of shrimp, lobsters, crabs, clams, oysters, scallops, and others in North America is only slightly exceeded by the annual harvest of finfish. More than 90,000 tons of honey is produced in the United States annually, and world estimates are over 800,000 tons.

Although plants provide most of the chemicals used to manufacture medicines, invertebrates have often proved important in fighting human disease and generating knowledge essential to many medical discoveries. Reflecting the medical potential of the spineless kingdom, some five thousand marine invertebrates have been identified as possessing cancer-fighting properties, and numerous human illnesses may be amenable to the biochemical properties of many invertebrate species. The scientific value of invertebrates has been enormous as a consequence of their huge numbers, incredible diversity, small size, and short regenerative cycles. Many current theories of taxonomy and evolution derive from studying invertebrates, for example, and an entirely new field, sociobiology, has resulted from studying mainly social insects. The experimental use of a single invertebrate family, the *Drosophila* fruit fly, has produced major advances in the understanding of genetics and heredity.

Invertebrates have also been the subject of art, fashion, and decoration. We decorate ourselves with silk, pearls, shells, and coral. Butterflies have been raised in recent years for decorative purposes—in fact, they represent an established industry in parts of Asia and South America. Musical composition has been inspired by invertebrates, and insects and spiders have figured prominently in language, myth, story, and symbol. Invertebrates have been the source of spiritual reflection—the ancient Egyptians worshiped scarab beetles—and there is a mythic link in many cultures between butterflies and the magical metamorphosis of a humble creature into one of the most beautiful animals endowed with the gift of flight.

Despite all these invertebrate values, a critical question remains: does the average person recognize or appreciate the positive virtues of these creatures? How, in fact, do people view invertebrates, especially insects and spiders? What do people know about these animals? How much do they support their conservation and right to exist?

Our study of attitudes toward invertebrates and knowledge of them focused on the general public, farmers, scientists, and members of environmental organizations. This research reveals that most people's knowledge of invertebrates appears to be quite limited. Only half of the public recognized invertebrates as lacking a backbone, few had any idea how many insect species exist, only a small number could distinguish one major invertebrate group from another, and many people confused invertebrates with various vertebrate species. For example, caterpillars were viewed by many as worms, the snail darter (an endangered fish) as a butterfly, the octopus as a type of fish, snails as more like turtles than spiders, spiders as a kind of insect. The least knowledge of invertebrates concerned matters of population status, animal behavior, and taxonomy. Greater knowledge was revealed toward pragmatically significant invertebrates such as insects in the garden, in agriculture, or related to human injury and disease. People knew the most about butterflies, perhaps because, like birds, these creatures are diurnal, colorful, and fly. They knew the least about termites, cockroaches, and beetles.

Few people seem even remotely aware of the tremendous number and variety of invertebrate species, especially insects. In fact, another study revealed that few Americans had ever heard of biological diversity and its current widespread loss, and none cited this loss as one of the most significant environmental problems facing the world today (although 15 percent did cite the closely related issues of deforestation and species extinction as critical problems). Even half the members of leading environmental organizations admitted never having heard of the current loss of biodiversity. The great majority of Americans appear to be unaware of the projected extinction of more than 10,000 invertebrate species every year.[27] Most invertebrate species officially listed in the United States as in danger of extinction (the Oklawaha sponge, Holsinger's groundwater planarian, Nickjack cave isopod, Kauai cave wolf spider, Keys scaly cricket, American burying beetle, North Platte montane butterfly, to mention just a few) represent for the great majority of Americans a rogue's gallery of obscure creatures with esoteric names and little apparent value.

Most of the general public and farmers view invertebrates with indifference, aversion, and disdain. Most in our study expressed dislike of bugs, beetles, ants, crabs, spiders, ticks, and cockroaches; a strong aversion to insects in the home; extreme dislike of biting and stinging insects; a desire to eliminate mosquitoes, cockroaches, spiders, fleas, and moths altogether. A large majority viewed invertebrates as lacking in the ability to experience pain, suffering, or consciousness. Most viewed invertebrates as incapable of feeling or emotion, affect or individuality, intellect or rational decision making. Few supported making significant expenditures or economic sacrifices on behalf of protecting endangered invertebrates.

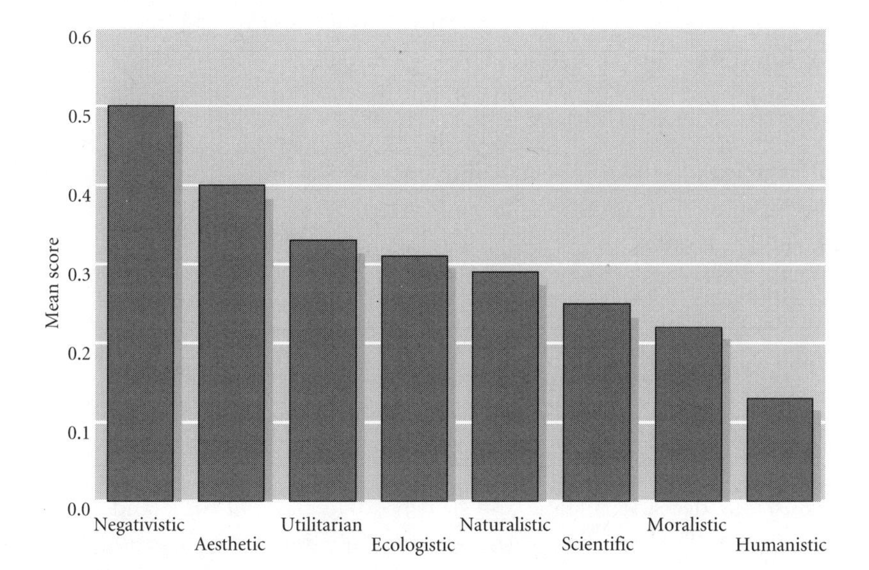

Figure 19. Attitudes Toward Invertebrates

We also constructed scales measuring people's basic attitudes toward invertebrates. As Figure 19 reveals, the prevalent view of invertebrates was a negativistic value of fear, dislike, and indifference. The least frequently encountered values were the humanistic and moralistic—nearly the opposite of people's attitudes toward vertebrate animals. Most people viewed invertebrates as creatures lacking in emotional and ethical value. These perceptions constitute a powerful basis for a view of invertebrates as largely worthless animals undeserving of protection.

The exception appeared to be invertebrates possessing unusual aesthetic, utilitarian, or naturalistic value. More positive attitudes were expressed, for example, toward butterflies, ladybugs, bees, shrimp, and "shells." The butterfly's beauty made it one of the best-liked invertebrates, and endangered butterflies were among the few insects to garner substantial public support for their protection. The practical value of bees probably constituted an important basis for their relative appeal, despite widespread aversion to stinging invertebrates. In these rare cases, people tended to put aside their general dislike of invertebrates in favor of admiration and appreciation, though rarely affection.

There was significant variation, too, between the general public, scientists, and members of environmental organizations. Scientists and, to a lesser extent, environmental group members expressed more favorable and sympathetic attitudes toward invertebrates. Scientists especially viewed insects as ecologically and pragmatically important and deserving of respect and protection. Environmental group members revealed generally positive attitudes, although to a far lesser extent than their corresponding views of vertebrate species. Farmers generally expressed antagonistic attitudes toward invertebrates, especially viewing insects as competitors needing suppression and elimination. Of all the groups, however, only a majority of farmers viewed bees as capable of conscious decision making and held positive views of these and other agriculturally beneficial invertebrates. Among demographic groups, the college-educated and, to a lesser extent, young persons revealed greater knowledge, appreciation, and concern for invertebrates and their conservation. Education again exerted a powerful impact on attitudes toward animals, independent of personal experiences involving these creatures. Antagonistic views of invertebrates emerged among the elderly, the less educated, and women in the study.

What is the basis for such pronounced feelings of dislike and indifference toward most invertebrates, particularly insects and spiders? Why do so many people seem unaware of the ecological and practical values of these creatures? Why do most view these animals with repulsion and even disgust?[28] One obvious reason for aversion to invertebrates stems from their widely assumed threat to human physical and material well-being. To many people, the modern conquest of hunger and disease has been achieved by eliminating and suppressing other organisms, particularly invertebrates. The spread of the human species across the planet has been facilitated by the mass production of a small number of agricultural crops and livestock. These monocultures have been made possible by an unrelenting war against invertebrates, especially insects. Thus the conditions of industrial-agricultural society may foster an adversarial perspective toward much of the invertebrate world. Many people also associate invertebrates with human illness. Modern theories of health treat many invertebrates as vectors of disease. An entire medical specialty, parasitology, has been constructed around the role of invertebrates in disease transmission, including the role of insects in spreading epidemics.

Perhaps humans have also developed a biological tendency to avoid certain invertebrates—particularly biting and stinging insects and spiders. Perhaps an aversion to certain arthropods conferred evolutionary advantages, becoming more preponderant over time in the human population. If this innate tendency does exist, one would expect it to become manifest with only slight provocation and, once stimulated, tend to remain. A number of studies have demonstrated this inclination—for example, exposure to spiders provokes withdrawal responses among many higher primates, including humans, even in the absence of an overt threat. This aversion to certain invertebrates may be evident, if only facetiously, in the views of the following "animal lover":

> Now, while I can't say that I exactly *love* beetles and grubs, I am literally the person who wouldn't hurt a fly, and have been known to amuse my family by running around with cardboard and jar to put our little segmented brethren out the screen door to live on. Nevertheless, and this is a big never and a big less, I have never been able to extend my feeling for the unity of all living things to the cockroach. Respect, yes—who could not respect these mortal enemies who survive combat (and everything else) to go forth into your blender mechanism and (alas!) multiply? But otherwise, I have no friendly feelings at all. Unlike other creatures, even other beetles, they seem to have no natural enemies (except me) and no obvious place in the great chain of being. Are you telling me that over time they break down my kitchen cabinet into soil and that is why they are here on earth? *Charlotte's Web* did wonders for spiders, but even *Archie and Mehitabel* could not warm my heart toward cockroaches.[29]

Perhaps people have become alienated by creatures fundamentally unlike themselves; perhaps they feel threatened, even offended, by the strangeness and otherworldliness of invertebrates. These creatures represent radically different solutions to the challenge of survival, having evolved in many unusual and seemingly alien ways. Their small size, odd body shapes, remarkable reproductive capacities, and apparent absence of a mental life represent characteristics very different from the human species. At the very least, these creatures seem exotic; at the worst, their oddness poses a threat to what is thought to be normal, acceptable, even permissible. For many people these organisms seem incomprehensible and incapable of evoking compassion. Their strangeness may stir sentiments of mystery and curiosity in some, but for most it prompts the urge to reject, even destroy.

Another reason for people's dislike of invertebrates may stem from their extraordinary numbers. A single bee hive may contain tens of thousands of animals, a single ant colony millions of creatures, an acre of soil more insects than entire nations, the number of known beetle species a third of a million or more. The psychologist James Hillman suggests that this multiplicity may constitute a fundamental challenge to human assumptions about the importance of individuality and selfhood.[30] These numbers threaten deeply held notions of the significance of a single life or an individual person's identity. As Hillman remarks: "Imagining insects numerically threatens the individualized fantasy of a unique and unitary human being. Their very numbers indicate insignificance of us as individuals."[31]

Hillman's training as a psychologist may have made him especially aware of the common association of invertebrates, especially insects and spiders, with notions of mindlessness and madness. Most people view invertebrates as lacking a mental life, emotion, or intellectual consciousness, all conditions of normality and moral standing in human society. It should not be surprising, then, that many metaphors of irrationality and insanity evoke images of insects and other invertebrates. As Hillman indicates: "Bug-eyed, spidery, worm, roach, blood-sucker, louse, going buggy, locked up in the bughouse—these are all terms of contempt supposedly characterizing inhuman traits. . . . To become an insect is to become a mindless creature without the warm blood of feeling."

We humans might also be frustrated by our frequent inability to master and control invertebrates, despite our presumptions regarding the omnipotence of modern science and technology. Although most vertebrates flee in panic when they encounter humans, many if not most insects and spiders appear oblivious, sometimes disdainful, of human presence. Invertebrates routinely take residence in our households, their aura of autonomy challenging our assumptions of preeminence even within the human home. The ubiquity of invertebrates, even the most feared, is suggested by the observation that most people will never live more than four feet from a spider.

Despite these reasons for human antipathy toward invertebrates, not all arouse such anxiety. Consider butterflies and bees. The aesthetic appeal of butterflies, and the utility and apparent intelligence of bees, have often endeared these creatures to people. A more respectful attitude toward invertebrates could perhaps be encouraged by making butterflies, bees, and other positively perceived invertebrates the moral equivalent of charismatic vertebrates like the wolf or whale. Deep-seated and perhaps innate aversion to invertebrates may prevent many people from ever developing strong affection for these animals. A more compelling delineation of their many contributions to human welfare, however, might counteract this hostility and instill a greater measure of respect for the wonder and diversity of these remarkable creatures.

Conclusion

This chapter has explored the many ways people attach value to varying species of life. Preferences were found to depend on elements of human relationship, use, and understanding. Attitudes toward certain species reflect the influence of basic values, a species' physical and behavioral characteristics, human knowledge and understanding, and past and present interactions. Attitudes toward wolves, whales, and invertebrates illustrate the often complex and dynamic character of human perceptions of the natural world.

Animals provide humans with food for thought as much as for the body. The wolf, whale, invertebrates, and countless other creatures offer people the means to project their fears, needs, illusions, and affections. The human ability to achieve feelings of familiarity, confidence, and well-being remains intimately dependent on highly varied and subtle relationships with the diversity of life. This sense of affiliation starts with creatures much like ourselves, but eventually it must extend to the remote world of "lower" plants and animals. Human respect for life, as Hillman observes, should include not just animals "in their splendor—the horned stag, the yellow lion and the great bear, or even old faithful 'spot'—but also those we fear the worst—the bugs."

CHAPTER 6

Culture

People's values of nature and living diversity have largely been explored from the perspective of American society. The views of other nations are important, however, if we are to understand the significance of cultural differences in perceptions of nature. This chapter partially redresses this imbalance by examining values of nature in other countries and cultures. After comparing basic views of nature in Western and Eastern societies, we will review some research conducted in Japan and Germany using methods similar to those we employed in the United States. Although this discussion will give us insight into the influence of culture on perspectives of nature and living diversity, particularly variations among Western and Eastern societies, this examination still focuses on modern industrial societies. To address views in a Third World culture, we will focus on the results of a comparable study in Botswana, a country with a population still pursuing a largely nonindustrial and, in some rare instances, hunter-gatherer way of life.

The role of culture in shaping values of nature confronts a basic question: do perspectives of the natural world constitute universal as opposed to relative expressions of the human condition? Do all people share certain fundamental values of nature independent of cultural and historical circumstance? Or do people construct unique and distinctive conceptions of the natural world depending on their economic, political, and social conditions? To phrase this question in another way: can cultures fabricate their own perspectives of nature and life (any value configuration being as intrinsically worthwhile as another)? Or are there only a limited number of ways people can value the living world in a healthy, functional, and sustainable manner?[1]

This book has a bias: that the basic values of living diversity have an inherent and evolutionary basis characteristic of all people independent of culture and history. These fundamental conceptions, however, represent weak biological dispositions shaped by learning and experience. The content, direction, intensity, and functional expression of these values depend on economics, politics, and other cultural conditions. The basic values of living diversity may remain constant, but culture modifies their form, content, and occurrence. Even so, a culture's long-term physical, emotional, and intellectual adaptability depends on the satisfactory expression of these values. To equate diversity of values with an infinite cultural capacity to construct unique conceptions of the natural world is to confuse content with underlying structure—ignoring how perceptions of nature orbit about a restricted core of inherent values of nature and living diversity.

Eastern and Western Perspectives

Despite this universality, there is remarkable plasticity in what a society views as legitimate feelings, beliefs, and behavior toward nature and wildlife. One of the most striking occurrences is the contrast between Western (Judeo-Christian) and Eastern (Buddhist-Hindu) societies. This difference is explored here by reviewing the scholarly literature, followed by results from studies in Japan, the United States, and Germany, the three most economically advanced Eastern and Western nations in the world today.

The historian Lynn White wrote an especially important essay in *Science* some three decades ago on Western conceptions of nature.[2] He argued that prevailing Western attitudes toward the natural world are rooted in Judeo-Christian religious beliefs, which have become closely associated with the development of Western science and technology. White contends that the West views humans as fundamentally different and superior to the rest of creation. According to the Western outlook, people alone possess the capacities for reason, moral choice, and spiritual transcendence of the physical world. This perspective, then, emphasizes a fundamental duality separating people from nature. Human uniqueness is seen as the consequence of a single governing force and an all-knowing God who conceived people in his image. Humans alone among all creation can attain freedom and release from the physical constraints of the material world—provided they adhere to the moral dictates of a world apart from nature. According to White, the belief in human uniqueness and superiority has encouraged both a contempt of nature and a desire to exercise mastery over it. Nature and nonhuman life are denied sacred status or the capacity for reason and selfhood—views White characterizes as representative of a rejected "pagan animism." Humans alone possess ethical standing, spiritual potential, and intrinsic value.

This prevailing Western perspective views nature as so much inanimate clay awaiting a higher transformation based on empirical knowledge and the application of technology. As the natural world has no feelings and selfhood, people can exploit, use, and control it in a "mood of indifference to the feelings of natural objects." From the Western viewpoint, "a tree [represents] no more than a physical fact." The natural world exists to serve human purposes, and the worth of nonhuman life is measured by its practical value. The subjugation of nature can be pursued, therefore, without guilt or inhibition. According to the cultural historian Keith Thomas, this Western outlook reflects a "breathtakingly anthropocentric" view: "Man's authority over the world was virtually unlimited. Animals [and nature] had no rights."[3]

Both science and technology flourished in this intellectual climate. Increasing material affluence and technical control over nature seemed to corroborate the Western assumption of progress contingent on humans transforming and dominating the natural world. The ultimate triumph of

Western civilization would allow people to transcend their natural and biological roots altogether. As Thomas observes: "The purpose of science was to restore to man dominion over creation. . . . [Western] civilization and science [became] synonymous with the conquest of nature. . . . The whole purpose of science was that [nature] may be mastered, managed, and used in the service of human life." This absence of restraint in exploiting nature assumed extraordinary dimensions of biological destructiveness when it converged with modern capitalism, science, and technology.

This depiction of Western conceptions of nature contrasts sharply with an idealized Eastern (Hindu-Buddhist) perspective of the natural world. Traditional Eastern views have been described as emphasizing a oneness that connects humans with all of creation—in contrast to the Western dichotomy separating people from the natural world.[4] The Eastern view is said to regard all living creatures as permeated with a similar life force, a fundamental kinship connecting all life in endless cycles of transformation and relationship. All creatures share a fundamentally similar experience, each striving after peace, harmony, and grace. All life, humans included, is thought to cohabit an analogous field of consciousness. People must respect and revere all living creatures, exercising kindness, practicing compassion, and avoiding harm to nonhuman life. Coexistence, rather than conquest, emerges in this view as a hallmark of Eastern thought.

Given the seminal significance of Lynn White's essay on Western perspectives of nature, a Japanese historian, Hiroyuki Watanabe,[5] published another *Science* paper five years later describing Eastern conceptions of nature and incorporating many of the assumptions just described. In contrast to White's portrayal of the West's antagonistic attitudes toward nature, Watanabe depicted an Eastern striving after harmony, respect, and peace with the natural world. He portrayed a traditional Japanese and, more broadly, Eastern "love of nature resulting in an . . . appreciation of the beauty of nature [and] of man . . . considered a part of nature, and the art of living in harmony with nature [being] the wisdom of life." Other scholars have similarly described aspects of this presumed Eastern affinity for the natural world, contrasting it with Western attitudes of superiority and arrogance toward nonhuman life.[6] Mariani, for instance, notes an Eastern "love and reverence for nature . . . a relationship based on feelings of awe and respect . . . in harmony with an understanding of nature's totality." Murota similarly suggests that Eastern thought views "nature as at once a blessing and friend [whereas] people in Western cultures . . . view nature as an object and . . . as an entity set in opposition to humankind."

Eastern reverence for nature has been linked to many traditional Japanese customs such as flower arranging, bonsai, the tea ceremony, rock gardening, certain poetry forms, and celebration of the seasons and beauty of the natural world. Saito suggests these practices have a certain perspective in common: "(1) nature as a friend of man, instead of servant to be used, enemy to be subdued, or an obstacle to be overcome; (2) no fundamental difference between man and nature; (3) the relationship between man and nature . . . as harmonious and unified."

Do these idealized depictions of Eastern and Western thought represent the attitudes of most people in these societies or just the views of a small educated elite? And do these traditional views persist in modern society, especially in industrial economies? Do these portrayals accurately describe the reality encountered in most contemporary Western and Eastern countries?

Some doubt emerges from the observation that many Eastern nations have inflicted considerable environmental damage, while many Western countries remain at the forefront of nature conservation and wildlife protection.[7] Modern Japan and China have been especially guilty of widespread habitat destruction, highly damaging harvesting practices, and overexploitation of many resources and wildlife species. Many Eastern nations consistently demonstrate environmental practices suggesting little respect, compassion, or concern for nature or living diversity. Many of the world's most ambitious environmental protection movements, by contrast, have occurred among Western nations, particularly the United States.[8] The presumably callous, arrogant, and destructive West has surprisingly revealed significant concern and even sacrifice on behalf of wildlife and nature protection.

This apparent gap between traditional representations of Eastern and Western concepts of nature and contemporary attitudes and practices prompted a comparative study of Japan and the United States.[9] This research provides some basis for examining contemporary Eastern and

Western views of nature, although the focus of this study was the world's two largest industrial economies. It might help first to review some socioeconomic and biogeographical characteristics of contemporary Japan and the United States before examining the results of this research.

The United States is a much larger country than Japan. America's population of some 260 million people is roughly twice the size of Japan's. The United States' much larger landmass, however, results in just 25 persons per square kilometer compared to 311 in Japan. In terms of arable land, there are 103 people per square kilometer in the United States compared to 2,256 in Japan. This difference stems largely from the concentration of the Japanese population along the narrow coastal plain. Some sections of Tokyo and Osaka have, as a consequence, population densities of more than 20,000 persons per square kilometer. The uneven Japanese population distribution is reflected in the fact that nearly two-thirds of the country, mainly the extensive mountainous areas, have less than 50 people per square kilometer.

This pattern of population size and distribution is related to very different biodiversity characteristics in the two countries. The United States' much larger area, lower population density, and relatively recent large-scale human settlement has resulted in a rich and diverse biota. Despite significant destruction, the United States retains all its large carnivores—often the first creatures to disappear when confronted with intensive, long-term human settlement. Japan, by contrast, has many fewer species, and most of its large mammals have been eliminated. Despite the country's considerable human population and long history, however, Japan retains a surprisingly impressive degree of biological diversity. For example, Japan has more than twice the number of bird and mammal species, and four or five times the insect species, as the comparably sized Great Britain and Ireland combined. This relatively diverse biota stems from three characteristics of Japan: its uneven population distribution, with few people residing in the two-thirds of the country consisting of mountains; a latitudinal range from the boreal forests of Hokkaido to the subtropics of Okinawa (comparable in distance from Canada's maritime provinces to the Caribbean Sea); and an archipelago that includes some four thousand islands with a combined coastline of almost 33,000 kilometers, possibly the second longest in the world. This extensive geographical and ecological variation has produced an unusual plant and animal life, including a large number of endemic species found nowhere else on earth.

Japan and the United States constitute the most industrialized and economically prosperous countries in the world. Together they account for more than one-third of the planet's gross national product and nearly half of its trade in goods and services. The United States possesses the larger economy and is still the world's leading exporter, although to a lower per capita extent than Japan. Japan has been one of the world's most successful industrial trading nations. Despite this fact, the percentage of Japanese who still engage in agriculture remains three times as great as in the United States.

Our study of Japanese attitudes toward nature and wildlife employed methods and concepts similar to those used in the United States, though often modified to reflect varying cultural, historical, and biogeographical characteristics. The Japanese study included personal interviews with a random sample of the Japanese population, as well as in-depth discussions with fifty Japanese scholars and leaders recommended for their knowledge of Japanese culture and perspectives of nature.

Interesting similarities and differences emerged among Americans and Japanese in their values of nature and living diversity. As these views have already been described for the United States, this discussion will focus on Japanese attitudes. Table 2 summarizes the relative attitude scores in the two countries.

These results suggest that contemporary Japanese and Americans have similarly strong emotional preferences for single species and individual elements of the landscape. Relatively high utilitarian scores in both countries also reflect a strong pragmatic orientation to nature and animals. Moderately pronounced naturalistic scores suggest a similar Japanese and American degree of interest in experiencing particularly favored species and outdoor recreational settings. Both nations, especially Japan, had relatively high negativistic scores indicating indifference toward aspects of the natural environment lacking cultural, practical, or historical significance.

A number of striking variations also distinguish Japanese and

Table 2. Mean Scores of Japanese and American Values of Nature and Wildlife

RANK	UNITED STATES		JAPAN	
1	Humanistic	(.38)	Humanistic	(.37)
2	Moralistic	(.28)	Negativistic	(.31)
3	Negativistic	(.26)	Dominionistic	(.28)
4	Utilitarian	(.23)	Naturalistic	(.22)
5	Ecologistic	(.22)	Utilitarian	(.22)
6	Naturalistic	(.20)	Moralistic	(.18)
7	Dominionistic	(.13)	Ecologistic	(.15)
	Knowledge	(.53)	Knowledge	(.48)

American attitudes toward nature and wildlife. Significantly higher dominionistic scores in Japan reflect a much greater stress on controlling nature—especially when certain species or landscapes have unusual aesthetic or emotional appeal. Much lower absolute and relative ecologistic and moralistic scores in Japan suggest far less concern than in the United States regarding the ethical treatment of animals, protection of natural habitats, or interest in ecological structure and function. Significantly lower knowledge scores in Japan indicate less factual understanding of wildlife and natural process than in the United States.

These statistical findings are complemented in many ways by the results of the in-depth interviews with various Japanese scholars, leaders, and environmentalists. These informants describe an especially pronounced Japanese emotional and aesthetic appreciation for the natural environment. They report, however, that this affection tends to focus on only a restricted range of favored creatures and landscapes and even then in a typically idealized fashion. Americans emphasize culturally familiar and aesthetically appealing creatures, too, but the Japanese perspective appears to be far narrower and often involves widespread indifference toward aspects of the natural world beyond a few favored species and landscapes. Most Japanese—in contrast to significant segments of the American population—express little interest in wild or pristine nature.

The Japanese leaders and scholars also describe a tendency among their countrymen to want to improve, enhance, and transform favored aspects of the natural world. There is a common desire, they say, to create or cultivate an ideal in nature, often requiring the isolation and exaggeration of especially important features of natural objects. The objective is to achieve beauty or perfection in nature through its considerable alteration and contrivance. There is an assumption, then, that properly appreciating nature requires manipulation, control, and cultivation—an aesthetic re-creation of preferred natural elements in the hope of isolating and most favorably expressing their centrally valued features.

The positive experience of nature and living diversity for most Japanese usually occurs under confined and artificial circumstances. Traditional nature appreciation activities—bonsai, haiku, flower arranging, the tea ceremony, rock gardening—reflect a refined appreciation of nature, even at times its veneration, but also a belief that wildness requires the creative hand and eye of humans to achieve its perfection. As one Japanese scholar remarked, these "nature-loving" traditions seem "more like having a pet with no idea of the basis of production, no understanding of these natural objects in a complete life cycle or in an ecological sense." Control and contrivance are necessary for effectively celebrating nature's hidden essence and beauty. Appreciating the natural world means cultivating its core value through adherence to strict rules and techniques.

This Japanese emotional and aesthetic affinity for nature often lacks an ecological understanding or particular interest in natural functioning. Indifference toward unmanaged species and landscapes prevails. Biological diversity beyond a few species and natural settings tends to be viewed with apathy and sometimes hostility. Japanese appreciation of nature and wildlife rarely includes much concern for ecological process, wildness, or more than a few favored species. A "love" of nature seems evident—but more a focus on altered than pristine nature, as well as an experience of living diversity requiring human will and mental re-creation to achieve its preferred expression. This appreciation often seems highly abstract and symbolic; an interest in nature from a safe and controlled distance. As one scholar remarked, most Japanese wish "to go to the edge of the forest, to view nature from across the river, to see natural beauty from a mountaintop, but rarely to enter into or immerse oneself in wildness or the ecological understanding of natural settings."

We encountered a Japanese affinity for nature, but largely an artistic and symbolic rendering of its original or primitive state. As one informant suggested, the Japanese wish to isolate favored aspects of the natural world and then "freeze and put walls around [it] . . . stealing aspects of nature and creating an art form around it." This refined appreciation involves a concept of nature's perfection dependent on following strict rules for seeing and experiencing nature's underlying harmony and beauty. Attaining this ideal demands human intervention and transformation of the living world. As Alan Graphard suggests: "What has been termed the Japanese love of nature is actually the Japanese love of cultural transformations . . . of a world that, if left alone, simply decays."[10]

Most Japanese did express an interest in experiencing and appreciating nature, but usually nature unburdened by its threatening or unmanaged character. This desire reflects an attraction to nature as a means of transcending the imperfections of the everyday world. The naturalistic experience allows one the opportunity for both escape and cultural refinement. As Saito remarks: "Nature is not lived or respected for its own sake but because it allows one to escape. . . . This appreciation of nature not only implies an anthropocentric attitude toward nature, but . . . suggests an ineffectiveness in generating an ethically desirable justification for protecting nature."[11]

This lack of interest in wild nature and ecological process encourages limited support for wildlife conservation and protection. These attitudes are reflected in the previously reported low ecologistic and moralistic scores of the Japanese. Limited Japanese concern for nature conservation extends to a general reluctance to grant ethical rights to animals. Other studies have similarly revealed that only a small proportion of the Japanese public evinces much interest in environmental issues or joins conservation organizations. A United Nations survey of fourteen countries found the Japanese last in awareness and concern for environmental problems. In another multinational study, most Japanese expressed a preference for interacting with nature in controlled and safe circumstances rather than confronting wilderness or pristine nature. Moreover, far fewer Japanese than Americans reported participating in outdoor activities such as hiking, camping, birding, fishing, or hunting. The Japanese did, however, report as much involvement as Americans in visiting zoos and watching nature-related television programs and films.

A Japanese reluctance to intervene actively in protecting and managing wildlife appears to reflect a belief that nature is beyond the grasp of humans to understand and control. Most Japanese tend to view people as fairly inconsequential before powerful natural forces. This presumption of human insignificance has produced an admirable sense of humility, but it has also encouraged a certain passivity among many Japanese regarding the human ability to conserve the natural environment. The scholar Watanabe explains this Japanese lack of conservation interest in this way: "If wild nature is beyond the human ability to control or grasp, if people are fundamentally inconsequential before an all powerful nature, then one can hardly imagine people as being able to exert much harm or regulation over nature."[12]

These Japanese ecologistic and moralistic values are further reflected in little difference among varying age and education groups on these scales. Not only were ecologistic and moralistic values rarely encountered among the Japanese public as a whole, but these perspectives appeared no more evident among better-educated and younger Japanese. This result contrasts greatly with that found in the United States and Germany, where college-educated and young adults reveal pronounced moralistic and ecologistic concerns for protecting wildlife and preserving ecological processes.[13]

Japanese perspectives of nature have been examined here to elucidate more general perspectives of living diversity in Eastern culture. Contemporary Japan, however, is a highly industrialized nation and perhaps no longer representative of traditional Eastern thought. Modern Japan might be more appropriately described as a nation that has adopted a Western paradigm and its related stress on science, technology, and control of the natural world. Some historians have suggested that predominantly Western perspectives became paramount in Japan starting with the Meiji period, two or three centuries ago, when Japan's leadership concluded that its survival required replacing prevailing Eastern culture with a Western model of economic and social development. Watanabe suggests that traditional Japanese affinities for nature have "been rapidly fading [ever] since this hasty introduction of modern science and technology." This presumption implies that environmental destruction in modern Japan and other Eastern nations largely reflects the Westernization of these countries. Baird Callicott, for example, contends that because of the "intellectual colonization of the East by the West . . . all Asian environmental ills . . . are either directly caused by Western technology . . . or aggravated by it."[14]

Westernization may indeed help to explain the paradox of contemporary Japan, where a strong tradition of nature appreciation appears to coexist along with widespread attitudes of indifference toward living diversity and its conservation. Many Japanese environmental values and beliefs, however, appear quite consistent with traditional Eastern views of nature. Eastern perspectives, for example, tend to be highly respectful and appreciative of the natural world, yet they also appear to be idealistic, abstract, and rarely associated with a detailed concern for the actual biological and ecological functioning of natural systems. Traditional Eastern attitudes also provide little explicit support for protecting or managing wilderness or living diversity, beyond a vague and indiscriminate covenant not to inflict suffering and to be compassionate to other creatures. Eastern attitudes further appear to encourage passivity and even fatalism toward an all-powerful nature viewed as beyond the capacity of humans to control, understand, or meaningfully manage. These traditional Eastern perspectives seem quite consistent with patterns of thought encountered among contemporary Japanese.

Based on these studies, it would appear that neither East nor West possesses an intrinsically superior or preferable conception of the natural world. Each society embodies both benign and antagonistic attitudes toward nature and wildlife. Before drawing any conclusions, however, let us examine prevailing values of nature in another Western nation, Germany, which along with the United States and Japan constitutes the third-largest industrial economy in the world today.

Germany

The study of German perspectives of nature and living diversity employed concepts and methods similar to those used in the United States and Japan. Dr. Wolfgang Schulz, a German scientist, conducted this research.[15] Before reviewing the results of this study, however, we should discuss some socioeconomic and biophysical characteristics of Germany.

Germany's population of approximately 70 million people is considerably less than that of Japan or the United States. Germany's population is roughly thirty times that of Oregon—an area approximately half the size of Germany. Germany's population density is approximately 250 people per square kilometer and, in relation to arable land, some 820 persons per square kilometer, a figure less than Japan but more than the United States. Biological diversity in Germany is limited compared with either Japan or the United States. Germany's long history of intensive agriculture, forest clearing, and urbanization, as well as its relatively recent glaciation, have produced a comparatively homogeneous landscape, relatively few plant and animal species, and, for all practical purposes, no remaining wilderness. Germany's limited biological variability is illustrated by the fact it has only five native conifers and thirty native hardwood trees, far fewer than found in either the United States or Japan. Germany also contains less than half the number of animal species encountered in either the United States or Japan.

Germany has simplified its forests by converting much of the country's native hardwoods to coniferous stands, as well as emphasizing the "production" of favored game species, particularly deer. Germany has, nonetheless, a long tradition of forest management resulting in nearly one-quarter of the country being retained in forestland. Germany also has a well-established tradition of wildlife management and appreciation, which, Arvid Nelson observes, has resulted in a strong cultural identification with wildlife, especially game animals: "The forest [is] deeply rooted in German tradition and myth continuing strongly today."[16]

Wolfgang Schulz's study of German attitudes toward wildlife reveals a strong generalized appreciation for wildlife similar to that encountered in the United States and far more pronounced than in Japan. The Germans and Americans expressed similarly strong interest in wilderness and wildlife, especially large carnivores. German views, however, tended to be more exaggerated, bordering on an idealization of nature and wildlife. Indeed, most Germans appeared to possess a romantic view of nature, stressing its ennobling qualities and the admirable virtues of wildlife and wilderness. To many Germans, the natural world assumed almost mythic proportions, signifying a simpler, more gracious, and harmonious life and time. This romantic idealization was also revealed in pronounced moralistic and ecologistic attitudes toward living diversity. The desire to protect wildlife was strongly evident among most Germans, particularly a view of animals as possessing rights independent of human interest or benefit. Many Germans expressed an unusual willingness to subordinate the practical needs of people to maintain pristine nature or protect wildlife.

The contrast between Germany and Japan is especially striking. Most Germans indicated a willingness to sacrifice human needs to defend the rights of animals and the natural world, unlike the Japanese tendency to assert control over nature unrelated to considerations of moral standing. Few Germans expressed a belief, common in Japan, that the realization of human needs necessitates subordinating nature to human will. The German ideal emphasized the inherent wisdom of a pristine natural world untouched by the presumably degrading impact of human intervention and constraint, in contrast to the Japanese emphasis on perfecting and transforming nature.

The overall configuration of German values more closely resembles America's than Japan's. Americans appear more like the Japanese, however, in their support for the practical use of living diversity and natural habitats. This utilitarian difference with Germany perhaps reflects America's relatively recent dependence on the harvesting and extraction of wild living resources for much of its wealth. The United States and Japan still maintain a strong emphasis on pragmatically utilizing nature, America mainly on land and Japan in a marine context. Both nations extensively use products derived from natural resources, and obtaining a living from the land and sea remains a well-established tradition, and myth, in both societies—especially the United States. By contrast, Germany long ago mastered and simplified its landscape, which was never particularly distinguished for its bounty of natural resources. Commerce and more recently industrialization have been Germany's primary sources of wealth.

German attitudes toward nature have, nonetheless, resulted in large tracts of preserved forestland. The country's long history of human settlement and high population density have led to a very managed landscape with few indigenous species remaining. Most of Germany's game species, for example, occur at artificially high numbers because of intensive management and habitat manipulation. Most Germans live apart from direct contact with nature or dependence on the land. Together these factors may have fostered a tendency to idealize wildlife and the natural world and the urge to see nature more as a source of inspiration and recreational enjoyment than as a provider of commodities for human use and profit.

Some Impressions

Considerable variation emerged in values of nature and its conservation in the United States, Japan, and Germany. This variability appears to be more a matter of degree, however, than any fundamental difference in each nation's basic perspectives of the living world. Indeed, all the values of living diversity appear in each society, though they may vary in content, intensity, and relative configuration. The biogeography, history, and culture of each nation have played a critical role in shaping their particular perspectives, but the basic value orientations remain the same. Each society appears to develop different strategies for expressing and realizing these perspectives of nature, but all cope to varying degrees with the same basic values.

Variations in concepts of nature among Eastern and Western societies appear to reflect the imprint culture places on the content and direction of the basic values. Each culture affects the potential of its people to achieve a rich and rewarding experience of nature and living diversity. Cultures can nurture the human craving for satisfying connections with the diversity of life, or they can frustrate this realization. One of the major traumas of the modern era, East or West, has been the emergence of an illusion that people no longer require intimate relationships with the living world to achieve lives replete in meaning and value.

Neither Western nor Eastern cultures appear to possess an intrinsically superior perspective of nature and living diversity. In each of these great civilizations, lessons for living in harmony and nurturing relation to the natural world can be discerned. From the East we may gain a heightened compassion and reverence for life, a sense of nature's oneness, and a profound tendency to live in harmony and balance with the natural world. From the West we can derive the desire to understand and directly experience natural process, the will to exercise environmental stewardship, and a belief in the human ability to manage the natural world wisely. A fusion of the best of each cultural tradition might very well lead to a new ethic of deep respect for the diversity of life, as well as a recognition that through our broadest affiliation with biotic complexity more humans can attain fully satisfying physical, emotional, intellectual, and spiritual lives.

Views in Nonindustrial Society: The Example of Botswana

This chapter has focused on perspectives of nature among Western/Judeo-Christian and Eastern/Buddhist-Hindu societies—a comparison derived largely from examining three of the world's leading industrial democracies. Conspicuously absent from this assessment were the views of nonindustrial, non-Western, and non-Eastern societies. Extraordinary cultural diversity, of course, exists among the earth's many nations. The consideration of variability of basic cultural conceptions of nature and wildlife throughout the world is clearly beyond the scope of this book. We will, therefore, only explore the views of nature in one Third World country, Botswana. This assessment draws from the work of a Nigerian researcher, Dr. Richard Mordi, who employed methods and concepts similar to those used in the United States, Germany, and Japan in his study of Botswana.[17]

Botswana is a southern African nation bounded by Zimbabwe in the east, Namibia in the west, Zambia to the north, and South Africa in the south. Botswana is a relatively large country with a landmass roughly comparable to California and Arkansas combined and substantially larger than either Japan or Germany. It is a mostly arid country: more than three-quarters of Botswana consists of the Kalahari Desert. Nonetheless, one of the earth's largest wetlands, the Okavango swamps, can be found in northern Botswana. This enormous wetland comprises the inland delta of the Okavango River, which empties into the Kalahari

Desert unlike the tendency of most large rivers to exit into the sea. Botswana has a human population of some 2 million people and a population density of just 10 people per square kilometer. The country's large landmass and low population density have allowed Botswana to set aside approximately 17 percent of the nation as protected area. The country remains endowed with spectacular concentrations of wildlife, particularly its large mammals and bird life.

Botswana is a largely nonindustrial country that enjoys a fairly high per capita income of roughly $1,000 per year. Some two-thirds of the population is literate. The nation's relative wealth and literacy derive from its abundant minerals, mainly diamonds, for Botswana is one of the world's largest producers of this precious stone. The country's relatively high standard of living and literacy rate have also contributed to a major shift of the population from the countryside to the urban areas, mainly the capital, Gaborone. Botswana's religious orientation remains roughly divided between Christianity and traditional animism. The dominant racial group is Bantu, although Botswana contains one of the largest remaining populations of hunter-gatherer Bushmen, some still practicing a traditional way of life.

Botswana remains a largely agricultural nation, focusing on the production of livestock because of its limited rainfall. Guaranteed beef markets in Europe have encouraged larger cattle herds, extensive groundwater extraction, and the creation of many permanent water holes. This increase in cattle numbers and water development has resulted in extensive overgrazing. The decline of Botswana's grasslands, as well as widespread fencing and increased wildlife damage control, have substantially reduced the country's wildlife. Another threat to Botswana's biological diversity is a plan to divert the waters of the Okavango Delta for agricultural and other development purposes. A rapid growth in wildlife ecotourism has concurrently occurred, particularly focusing on the Okavango Delta and Chobe River areas.

Richard Mordi's study found all the basic values of living diversity in Botswana—although, not surprisingly, to a different degree than occurred in the United States, Germany, or Japan. Moreover, Mordi noted one additional perspective of the natural world not previously encountered in the other studies. Specifically, he described a view of nature somewhat related to the moralistic perspective but possessing distinctive properties of its own. Mordi labeled this value theistic. This perspective views nature and wildlife as largely beyond the control and direct experience of humans. Its most distinguishing characteristic is a perception of the natural world as reflecting the will of supernatural forces or deities who govern the destiny of both humans and wildlife. The theistic value regards people as possessing little influence over the status of animals or the course of natural events. All living beings reflect processes and powers outside the human realm. People are largely impotent before these forces, lacking the capacity to regulate or manage the living world to any significant degree. The abundance of wildlife derives mostly from these transcendent and magical forces. An attitude of fatalism prevails since humans possess limited ability to intervene significantly in nature beyond a generalized appeal to supernatural forces for assistance and generosity. The fate of beasts and humans alike depends mainly on luck, destiny, or magical intervention.

Mordi found the theistic and utilitarian values to be the most dominant views of nature in Botswana. For most Botswanans, wildlife possessed largely practical or magical significance. Species with traditional totemic or commercial importance tended to be the most relevant and preferred. Considerable antagonism prevailed toward creatures whose subordination or even eradication pragmatically benefited people, particularly cattle production, or were the subject of strong superstitious dislike.

The widespread occurrence of the theistic value means that few Botswanans view efforts to promote the management or protection of wildlife as especially relevant or worthwhile. Since supernatural forces dictate the fate of wildlife, any attempt to modify this situation could risk inviting the retributive wrath of the deities. Human efforts to influence or substantially alter nature are viewed as likely to be ineffectual and possibly provocative. The theistic perspective suggests that people should accept the will of supernatural forces rather than chance offending these powers by arrogantly trying to abrogate their prerogatives.

Most Botswanans also expressed strong humanistic and aesthetic values toward the natural world. Certain wildlife possess great emotional and symbolic appeal, especially large, charismatic species. Many Botswanans further revealed a pronounced affection for pet animals.

These humanistic and aesthetic preferences appear especially among more affluent, better-educated, and urban Botswanans.

Except for a few favored species, most Botswanans expressed pronounced negativistic values toward wildlife and the natural environment. Botswanans tend to view most wild animals with indifference and often fear and hostility. Only a small minority revealed significant naturalistic interest in wildlife and natural habitats, and even a smaller number manifested much ecologistic concern, scientific interest, or factual knowledge of the natural world.

A striking exception to this general pattern of apathy, fatalism, and materialism toward nature and wildlife emerges among better-educated and urban Botswanans. These demographic groups revealed substantially greater intellectual and emotional concern for nature and living diversity. Recreational wildlife interest among better-educated and urban Botswanans tended to be relatively passive, however, focusing on wildlife on film and television or visiting zoos and highly managed parks.

Most Botswanans, therefore, viewed nature and wildlife with pronounced utilitarian, theistic, and negativistic values and revealed little ecological appreciation, knowledge, or moral concern for living diversity. Mordi suggests the combination of these attitudes constitutes a formidable barrier to biological conservation in Botswana. He concludes it would be difficult to obtain widespread support for nature protection in Botswana unless it was based on convincing economic arguments, such as the utilitarian benefits of ecotourism, game ranching, and other forms of wildlife utilization. Mordi predicts that unless there is a fundamental change in values toward the natural world, Botswana's rich and varied fauna will decline as the country becomes industrially developed and urbanized. More appreciative environmental attitudes among better-educated and urban Botswanans suggest that as Botswana prospers a broader constituency could emerge more strongly committed to conserving the nation's extraordinary biological heritage.

Hunter-Gatherer Society

Before concluding, we should consider another Botswanan cultural tradition toward nature and wildlife. As indicated, a small number of hunter-gatherer Bushmen still exist in Botswana, mainly residing in the

Kalahari Desert and Okavango swamp areas. As there is little information about Bushmen values of living diversity, this discussion relies on general impressions and findings reported for other hunter-gatherer peoples.[18]

Traditional hunter-gatherer societies typically maintain self-sufficiency in food production, deriving most of their sustenance from hunting wild animals or gathering wild plants. Hunter-gatherers tend to rely only marginally on market or cash economies or on the organized production of livestock or crops. Hunter-gatherers usually lack fixed territories, are highly mobile, have small populations, and possess little or no written language. They tend to employ simple technologies, possess few artifacts, mostly made of local materials, and maintain few continuous contacts with the modern world.

Various studies have suggested hunter-gatherer peoples evince a largely positive, even reverent, attitude toward nature and living diversity. A key feature of this veneration appears to be the belief in a profound sense of connection between people and the nonhuman world. All living beings, people included, are thought to share a common field of consciousness. The distinction between human society and the external world becomes blurred by the conviction that human experience merely represents another extension of forces prevalent in the natural world. This perspective typically elevates nonhuman life to a status roughly equivalent to that possessed by humans.

Hunter-gatherer peoples often assume that a spiritual kinship binds people with animals: both are subject to simultaneously controlling and reciprocal relationships. An affinity and respect for nonhuman life tends to be fostered by a view that Richard Nelson describes as imbuing "all of nature with spiritual powers. . . . All creatures, no matter how small and inconspicuous, carry the luminescence of power."[19] Humans and animals are seen as occupying and sharing a common social order. Mutual obligation, dependency, and reciprocity between people and wildlife prevail with binding duties and associated obligations. As in any society, only restrained and respectful patterns of behavior can reasonably assure harmonious relations.

This presumption of reciprocity between humans and nature tends to encourage an attitude of respect, even humility, toward the natural world. If humans fail to treat nonhumans with deference, people will be regarded with equivalent indifference. For most tribal hunter-gatherers, these assumptions frequently lead to prohibitions against gratuitous exploitation, as well as callous, cruel, or harmful conduct toward animals. Strong utilitarian values may prevail, but they tend to be inhibited by respect for other beings, who presumably share a rough kinship and equivalence, and by powerful fears of retaliation from the spirit of these creatures for abusive actions.

Norms against excessive wildlife exploitation tend to be motivated more by sentiments of affinity and reciprocal obligation than by ecologistic or moralistic values typical of modern rationalizations for protecting nature. For most hunter-gatherers, moderation and respect for the natural world derive from a profound emotional and spiritual identification with the living world, unrelated to a calculated empiricism or a particular desire to prevent pain being inflicted on other creatures. The typical hunter-gatherer assumes that avoidance of waste and excessive harm to nonhuman animals is as logical as preventing similar behavior from befalling one's own family, village, or tribe. Such hurtful actions not only injure the object of one's affection but invite retaliation. As Nelson suggests, most hunter-gatherers assume that "the environment is inhabited by watchful and potent beings. . . . Violations against nature can bring every sort of bad luck or personal harm. . . . People should do everything possible to prevent unnecessary suffering; they should never take more than they need or waste what has been given them."

Only a small number of Bushmen remain in Botswana; even fewer maintain a traditional hunter-gatherer way of life. The Bushmen, like most of the world's hunter-gatherers, have experienced significant declines during this century, for the destruction of human cultural diversity has roughly paralleled the demise of biological diversity. The decline of tribal cultures has also encouraged the disintegration of traditional patterns of resource restraint and respect for nature. Feelings of intimacy and kinship with wildlife have been replaced for many tribal and hunter-gatherer peoples by exploitative values associated with contemporary marketplace economies, modern technology, and commercialism. Many hunter-gatherers have responded to new cash incentives by engaging in excessive harvesting, succumbing to the temptation of overexploiting wildlife in exchange for receiving outside manufactured goods.[20] Frequently all that remains of traditional values of nonhuman life are lingering theistic beliefs in the inability of humans to control nature, but now dangerously associated with a reward system that often encourages excessive and unsustainable wildlife exploitation.

The modern nation-state, by severing traditional dependencies on living diversity, often encourages indifference toward a natural world that no longer seems particularly relevant. Tribal values of conserving and protecting nonhuman life are rendered spiritually inoperable, while new ecological and ethical foundations for sustaining nature have not yet emerged. Many developing nations with recent histories of tribal dependence on nature, like Botswana, consequently appear suspended between two worlds. Traditional epistemologies emphasizing a basic connection between humans and animals have lost much of their practical meaning, while new intellectual concepts of respect and affinity for nature have not yet fully and persuasively developed.

Conclusion

The cultural traditions examined in this chapter—Western, Eastern, industrial, developing, and hunter-gatherer—all possess powerful elements of appreciation and concern for nature and living diversity. Each tradition manifests elements of affinity and affection for nature, recognizing how the celebration of life can enrich the human body, mind, and spirit. Each has its limitations, however, imposed by its economic and historical orientation. An evolving ethic of respect for the living world can draw inspiration and guidance from the strengths inherent in each cultural tradition. From the West, one can derive a greater understanding of the ecological connections that bind life together in a vast matrix of interdependencies and relationships. From the East, one can encounter a compelling articulation of kindness and compassion for nonhuman life and a striving after harmony and balance with the natural world. From the tribal, one can discern how other organisms constitute parallel nations who, if one listens carefully and watches closely, can communicate a vast and enduring wisdom.

Part Three

Applications

CHAPTER 7

Endangered Species

THE BOOK began by outlining a framework of basic values of living diversity thought to originate in the biological character of the human species but subject to the formative influence of human experience, culture, and history. In Part Two we explored these value differences among varying demographic, activity, and cultural groups and in relation to different species. In Part Three, the final section, we consider the link between values of nature and the challenge of managing, conserving, and restoring living diversity.

This chapter considers the relation of people's attitudes toward living diversity and the problem of conserving endangered wildlife. Protecting and restoring endangered species represents a complex challenge. The multidimensional nature of the task is suggested by the policy framework depicted in Figure 20, developed by Tim Clark and myself.[1] As this diagram indicates, effective conservation policy, whether for endangered species or any other wildlife, requires the consideration of biophysical, socioeconomic, institutional-regulatory, and

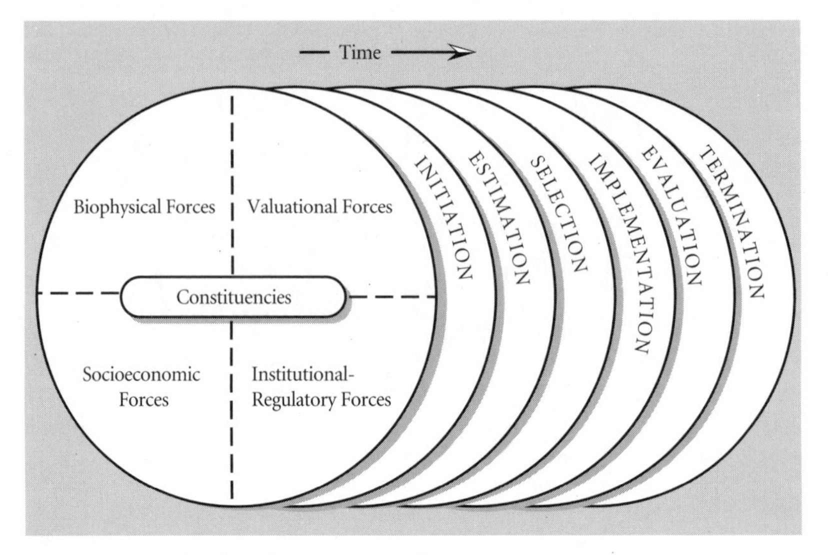

Figure 20. Wildlife Policy Framework

valuational forces. Moreover, these dimensions must be understood in relation to the competitive interactions of various stakeholders. And all these forces and interactions tend to change with time. Endangered species and wildlife policy should thus be regarded as a complex multidimensional, dynamic, and dialectical process, difficult to understand, and even harder to control and render more efficient and effective.

It might be helpful to explain this wildlife policy model in relation to endangered species conservation, starting with the multidimensional factors. Biophysical forces emphasize the role of biotic and abiotic processes in protecting and recovering imperiled species. Socioeconomic factors stress the significance of varying power, authority, and property relationships within communities and societies in conserving endangered wildlife. The institutional-regulatory dimension emphasizes the role of politics, law, and organizational behavior, particularly among government agencies, in designing and implementing effective species protection and recovery programs. The valuational dimension, the focus of this chapter and book, stresses the importance of human judgments and perceptions in conserving and restoring endangered wildlife.

Endangered species policies are also formulated in a context of competing stakeholders, each claiming certain interests in how biotic resources and their habitats are used and managed. Ignoring these group relationships—or responding to them insufficiently—usually results in ineffective, contentious, and unacceptable endangered species policies. Resource policies also shift with time, as the policy analysts Garry Brewer and Robert DeLeon have found.[2] Endangered species programs tend to change from their initial formulation and estimation of varying options, to the selection of a particular course of action, to implementing these choices, to eventually evaluating the outcomes and changing or terminating the conservation efforts. Each policy stage presents different constraints and opportunities, and every endangered species program must respond appropriately to these temporal realities.

This chapter focuses on the importance of human values in endangered species policy. After reviewing the United States Endangered Species Act (ESA), arguably the most powerful and ambitious wildlife law ever enacted, we will look at three case studies: an endangered mammal, the black-footed ferret; a rare fish, the snail darter; and an imperiled bird, the palila. Each species' story will bring into sharp relief the complexity of factors associated with endangered species policy and protection, particularly the role of human values in the survival of these creatures.

Although the Endangered Species Act was enacted in 1973, earlier versions had been passed in 1966 and 1969. These versions, however, had several shortcomings: a restricted geographic and taxonomic focus, ineffective controls on harming endangered species and their habitats, and inadequate procedures for implementing various protection requirements.[3] Prior legislation had been developed earlier in the century, but these measures focused only on certain species or had very limited application. The Lacey Act of 1900, for example, America's first federal wildlife law, assisted in eliminating the commercial wildlife trade, which had driven many species such as the American bison and many shorebirds to the edge of extinction. This act only focused, however, on commercial causes of endangerment and provided little explicit emphasis on protecting endangered wildlife. Other important legislation of the time included the Bald Eagle Protection Act, the Northern Fur Seal Treaty, and international whaling agreements, but each focused on a single species or just a few species, typically creatures of economic or cultural significance.

Moreover, these laws had limited powers and depended on the discretionary commitment of varying government agencies for their effective implementation.

The Endangered Species Act of 1973 constituted an unprecedented exercise of legal ambition, innovation, and comprehensiveness. The ESA represented the most powerful declaration ever to preserve and protect wildlife, theoretically subordinating all other considerations to the imperative of preventing species from becoming extinct. Unlike previous efforts, the ESA declared all life forms, invertebrates included, anywhere in the world, as worthy of protection. The act further proclaimed that the government has a solemn duty and trust to protect these creatures on behalf of the American people, not just for their commercial and material value, but also because they represent irreplaceable ecological, scientific, recreational, aesthetic, and ethical values as well.

The ESA further recognized that species might be in danger of extinction at different levels of vulnerability requiring different strategies of management and protection: species in imminent danger of extinction; less seriously threatened species; endangered subspecies; distinct populations of an endangered species; species closely resembling an endangered species; and even, in recent years, species proposed for formal protection but not yet officially reviewed and listed. The ESA further emphasized the importance of protecting not only species from direct harm, but their habitats as well, and the need to designate lands and waters critical to the continued existence of a species. Another provision prohibited the "taking" of an endangered species, eventually interpreted by the courts to include not only direct harm or mortality inflicted on a species but also destruction or significant modification of its habitat.

The ESA required that all federal actions which might compromise a species' existence must be reviewed—and if these actions proved harmful and unavoidable, they must either be modified or eliminated altogether. The ESA additionally allowed an unprecedented degree of public involvement in proposing the listing of a species, participating in the review, and, most important of all, bringing court action against other parties, including the government, when the act appeared to be violated. The ESA declared among its many goals not just preventing the extinction of a species but the far more ambitious and positive goal of removing creatures from the endangered species list because they had recovered. Finally, the ESA included a provision for its reauthorization every five years—theoretically facilitating the act's refinement and improvement, but also opening the ESA to considerable political pressure and revision.

This cursory overview of the ESA's remarkable breadth and complexity has not even touched on the many legislative amendments and court interpretations that have substantially revised the act during the quarter century of its existence. Some of these changes and related issues will be raised in the cases to be examined. For now, the major point is the ESA's unusual degree of rationality, comprehensiveness, idealism, and scientific orientation, at least on paper. Despite the ESA's remarkably ambitious and visionary character, its implementation has been highly contentious and less sweeping than might have been expected. The ESA has been seriously attacked, particularly in recent years, not only by those who oppose vigorous endangered species protection, but also by those who favor it.

Opponents of the Endangered Species Act argue that it has had a crippling affect on legitimate socioeconomic interests, trampling individual liberties and property rights, and has been used as an antidevelopment tool often unrelated to the needs of imperiled wildlife.[4] These critics suggest the act has been a nightmare of government regulation and restriction, interfering with activities vital to the nation, and sometimes constituting an unwarranted intrusion on constitutionally guaranteed liberties. The seemingly inflexible nature of the ESA, and its tendency to place species protection above all other considerations, has been especially galling to these critics. They particularly object to the economic burden of species conservation falling on those whose activities or property rights are restricted, while the benefits of protection accrue largely to society as a whole. Finally, many have been outraged at the preemption of local and state rights by the exercise of expanded federal authority to protect endangered wildlife.

Advocates of vigorous endangered species conservation counter that the ESA has fallen far short of its protection goals. Some even claim the act no longer functions as a serious safety net against extinction.[5] Those who would like the ESA to be more expansively implemented believe the responsible government agencies, the U.S. Fish and Wildlife Service and the National Marine Fisheries Service, have been unduly influenced by powerful political and economic interests. Such critics assert these government agencies have largely transformed a prohibitive and unequivocal mandate into a discretionary act resulting in piecemeal and often ineffectual protection and recovery of endangered wildlife. They cite various performance deficiencies to support their case.[6] Some of these shortfalls include: a slow and costly listing process (averaging, until recently, some fifty species per year at a cost of $75,000 per species); a large number of "candidate" species proposed for listing but not yet officially afforded protection (until recently, more than three thousand candidate species awaiting review); few species removed from the endangered species list for reasons of recovery (and a number withdrawn for reasons of extinction); a minority of species with formally approved recovery plans (and even fewer with implemented plans); a minority of endangered species with officially designated critical habitat; a small fraction of proposed government activities (water projects, highway construction, oil and mining development) significantly modified, and a much smaller proportion eliminated, because of their presumably adverse impacts on endangered wildlife.

It is impossible to arrive here at definitive conclusions about these many issues. Both sides appear to have legitimate grievances. The protection of endangered species must occur in a just and democratic manner respectful of the economic needs and civil liberties of a nation's citizens and institutions. Within this libertarian framework, a diverse and abundant biota remains essential for an economically, culturally, and environmentally healthy and sustainable society. The ESA's effectiveness represents a critical component in halting the current hemorrhaging of life on earth. It is still this country's most powerful legal tool for confronting the contemporary biodiversity crisis.

Some suggestions will be offered toward the chapter's conclusion regarding how the ESA might be applied more effectively. For now, let us explore the complexities of endangered species implementation, particularly the role of human values, by examining three interesting and relevant case studies: the snail darter, the palila, and the black-footed ferret.

The Snail Darter

The snail darter was until recently a practically unknown and diminutive fish that fit comfortably into an adult's palm. Shortly after the Endangered Species Act's passage, this little creature starred in the earliest and perhaps most contentious test case of the act's broad and prohibitive application.[7] This famous controversy pitted the "little" fish against the "big" dam. The venue was Tennessee in the mid-1970s, along a small tributary of the Little Tennessee River, a site where the Tennessee Valley Authority proposed completing a dam on the Tellico River, the only known habitat of the snail darter. The ESA had recently been enacted, but most policymakers and the public assumed it focused on well-known species like the bald eagle, grizzly bear, whooping crane, or even arguably wolves: creatures, for the most part, whose imperiled status had stemmed largely from overexploitation and past persecution. The snail darter controversy suggested for perhaps the first time that the Endangered Species Act might actually require major economic sacrifices of popular projects on behalf of protecting obscure and seemingly worthless organisms. Moreover, this effort might conflict with and possibly triumph over well-established and powerful political interests.

Some background on the snail darter might be helpful. This diminutive fish is a two- to three-inch member of the perch family. Some 130 darter species exist in the Tennessee River system alone, nearly 90 in the state of Tennessee. Many darter species coevolved with equally unique freshwater mollusks, the snail darter's name connoting its dependence on these freshwater creatures for its survival. Both snail darters and their associated mollusks require cool, clean, highly oxygenated waters, characteristic of fast-flowing streams, with a gravel shoal streambed. Like most creatures who have evolved in relative isolation, snail darters are specialists requiring certain habitats, food sources, and reproductive areas for their survival.

Many freshwater fish species and mollusks have suffered gravely as a consequence of massive habitat destruction, modification, and alteration of America's mighty river systems including the Tennessee, Ohio, Mississippi, and others. The Tennessee River drainage had been particularly devastated, mainly because of its long association with one of the world's largest public works projects, the Tennessee Valley Authority (TVA).

The TVA was initially conceived during the depths of the Depression to promote economic revitalization of this highly impoverished region. Major activities of the TVA included large-scale generation of energy and the controlling of floods by building dams and other impoundments along the Tennessee River and its various tributaries. Over the years, the TVA seemed to have evolved into a public works agency largely intent on providing jobs, serving various political interests, and, like any good bureaucracy, ensuring its continued survival. The construction of the Tellico Dam conformed with all of these bureaucratic-maintenance, job-creating, and political objectives. Moreover, it focused on an area that already had many dams. Indeed, the site of the Tellico Dam involved one of the last fast-flowing streams in a region where most similar habitats had already been converted into still, deep, and, for most endemic darter and snail species, dead lakes.

The Tellico Dam's construction had been initially proposed in the 1930s, but the project was deferred at the time because so many other preferred sites existed. By the 1960s, most of these dams had been constructed. The Tellico project was revived, in many people's minds, to keep the TVA machinery running. Dam proponents initially claimed the project would produce recreational, navigational, and flood control benefits, as well as provide industrial development sites along the lake shoreline created by the dam. These claims were not especially convincing, though, for many other nearby dams already provided similar benefits. As the debate over halting the largely completed Tellico Dam evolved, TVA shifted its major argument to the project's supposed hydroelectric benefits—which, given the energy crisis mentality of the time, constituted a more politically compelling assertion.

But there were just as many reasons for not completing the dam. Prime agricultural land would be lost along the river corridor, as well as one of the area's few remaining sport fisheries associated with an unaltered stretch of fast-flowing stream. The area to be flooded also included the spiritual heartland of the Cherokee people. Most of all, from the perspective of this discussion, the Little Tennessee River was identified as the only known habitat at the time of the snail darter and various associated creatures. The snail darter's habitat appeared to be as much in danger as the species. The snail darter represented a kind of "miner's canary"—an indicator of the health of an increasingly rare ecosystem integral to the survival of an entire community of organisms. By creating the dam, the Tellico River would be shifted from a fast-flowing stream to a still lake. This would silt and eliminate the gravel beds essential to the snail darter and other creatures' feeding, reproductive, and associated survival requirements.

TVA conducted a cost-benefit analysis of the advantages and disadvantages of completing the project. Like most efforts at quantifying the value of economic development versus the preservation of an obscure species, the calculation was strained and distorted. TVA's calculus, not surprisingly, favored activities with a well-established marketplace value, distinctly minimizing the importance of benefits not amenable to economic expression. The cost-benefit analysis stressed the dollar value of land development, bridge construction, dam building, reservoir creation, flood control, hydroelectric energy, navigation and water supply, and recreation. Virtually no discernible economic return was assigned to the cultural importance of the region, the significance of an increasingly rare habitat, or the value of an endangered species like the snail darter or the even more "lowly" freshwater mollusks. Based on this analysis, TVA recommended completing the dam. Known economic returns clearly outweighed vague benefits of an unknown and inedible fish, local agricultural interests, a fast-flowing stream, or a Native American constituency more relevant to the past than to the present. An overwhelming bias existed toward activities measurable in dollar terms and associated with well-established human needs.

The Endangered Species Act became involved as an unyielding statement of the higher priority of protecting a listed species over all other considerations, especially those involving federal agencies like the Tennessee Valley Authority. The ESA allowed no weighing of the costs of the snail darter's survival against the socioeconomic benefits of the development project. Moreover, the ESA's unusual public access provisions allowed a citizen's group to bring legal action against the Tennessee Valley Authority to halt construction of the Tellico Dam. Politicians, interest groups, federal agencies, and much of the media became aroused by this attempt to stop a popular and powerful development agency in order to protect an obscure and economically insignificant fish. The snail darter represented, in effect, the first major test of the ESA's extraordinary objective of preserving unique life forms even when it meant restricting or blocking the activities of powerful interests. Zygmunt Platter, who successfully defended the snail darter case before the Supreme Court, suggested this dispute constituted a classic confrontation of new and barely emergent environmental values against deeply entrenched assumptions of social worth and economic importance.

The Supreme Court eventually upheld the protection of the snail darter as dictated by the Endangered Species Act, thereby halting the completion of the Tellico Dam. It refused to alter, through interpretation, the unequivocal language of the law which placed the existence of a listed endangered species above all other considerations. Chief Justice Warren Burger voiced his skepticism about preserving a species of dubious value over the demonstrated utility of an important development project, but he declared it was Congress's role to change the law rather than the courts to revise it through legal ruling.

Congress eventually did amend the ESA, largely in reaction to the Tellico Dam case. The Endangered Species Act's reauthorization in 1978 included a provision that in rare circumstances society's interests could be weighed and balanced against the protection of a listed species. The amendment created a special committee of high-level government officials who might be convened in cases where species protection conflicted with vital public interests. This "God Squad," as it has been called, has met fewer than a dozen times since the amendment was enacted, rarely finding that a jeopardizing activity should take precedence over the protection of an endangered species. The committee even concluded in the snail darter case that the benefits of the Tellico Dam did not outweigh the threat to the existence of this imperiled fish. Congress, however, through appropriations legislation, allocated funds to complete the dam. Moreover, President Carter refused to veto the water bill—despite his strong religious convictions, which apparently did not conflict with the presumed sanctity of another species' existence.

President Carter's views appear to be quite consistent with prevailing attitudes toward the snail darter and other obscure fish, as reflected in a finding from our national study cited earlier. As Figure 21 indicates, an overwhelming majority of Americans support water projects that result in significant human benefits such as hydroelectric energy, increased drinking supplies, or water for irrigation, even if the project results in the extinction of an obscure fish species like the snail darter. Moreover, all demographic groups endorse these water projects, including a majority of the college educated, young adults, and even members of environmental organizations. Only a minority of Americans, however, support less crucial water projects—like diverting water for cooling industrial plant machinery or creating a lake for recreational purposes—if it jeopardizes the existence of an unknown fish species.

Support for water projects imperiling the survival of aquatic life is especially distressing given that nearly one-third of the country's indigenous freshwater fish may be in danger of extinction, mainly due to significant habitat modifications. Water projects justified in the name of agriculture, energy, drinking water, and other human benefits have already resulted in major reductions in aquatic biodiversity and the elimination of many species. Public support for these projects may help explain why the General Accounting Office found that of some three

Figure 21. Attitudes Toward Water Projects That Endanger Fish Species

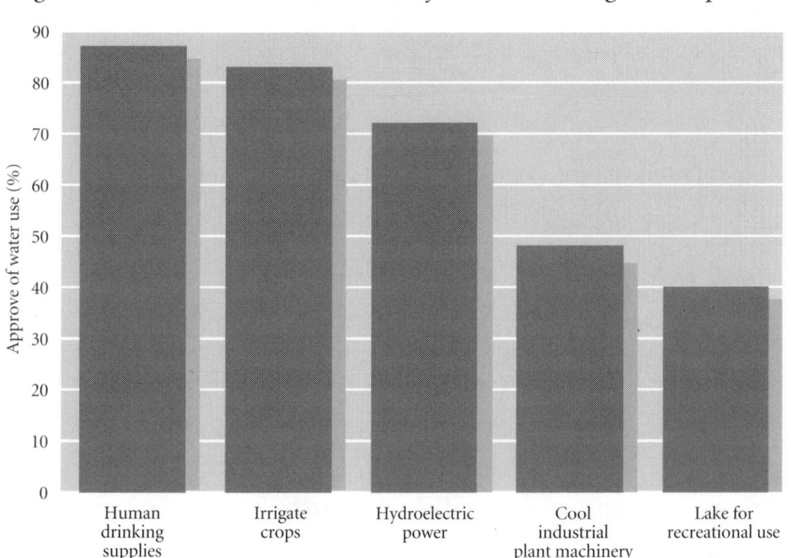

thousand proposed western water projects, not one had been halted because of presumed harm to a listed endangered or threatened species.

The snail darter case and the survey results indicate the scope of the challenge of trying to cultivate a more sympathetic attitude toward the conservation of obscure and seemingly useless endangered creatures. For most people, the construction of large dams and associated benefits represent another illustration of the success of modern technology and civilization. These projects rarely provoke misgivings that these successes might destroy countless creatures and their habitats.

Still, the snail darter case may one day be viewed as a profound turning point in the evolution of a new consciousness toward the preservation of all life. It forced many to confront for perhaps the first time the scope of the problem of species loss, and it elevated the plight of an obscure species to unprecedented heights in questioning powerful economic and political interests. It may even have expanded public awareness of habitat destruction as a fundamental cause of species endangerment and biological destruction. This case might just emerge as a celebratory moment when our society began altering the inertial flow of an enormous tide of destruction that had prevailed unimpeded for centuries.

The Palila

The Endangered Species Act and public values toward imperiled wildlife were also greatly affected by another relatively obscure creature, this time a Hawaiian bird called the palila.[8] The palila is a member of the honeycreeper family of birds, one of the most biologically unique group of animals in the world. The uniqueness of the honeycreepers stems from their having evolved on the Hawaiian Islands, the most geographically isolated landmass on the planet and islands that never have been connected to a continent. This isolation resulted in a proliferation of many unique species on Hawaii, dramatically represented by its unusual birdlife.

At the time of the islands' discovery by Captain Cook in 1778, Hawaii had perhaps eighty endemic bird species, including an estimated forty-seven honeycreeper species and subspecies, the latter having apparently evolved from a single founding animal that had colonized the islands by chance.[9] Hawaii's geographic isolation and the absence of land predators encouraged the remarkable evolutionary radiation of the honeycreepers, and many species became specialists exploiting a particular niche among the islands' lush variety of habitats and food sources. The honeycreepers developed an extraordinary variety of bills and body shapes. Indeed, the honeycreepers' evolution might have been a better illustration of Charles Darwin's theories of evolution than the finches he made so famous on the Galápagos Islands. Perhaps if Darwin had traveled to Hawaii much of the islands, unique plant and animal life might not have been lost during the two centuries since Captain Cook's pioneering voyage. Of Hawaii's eighty endemic bird species in 1778, twenty-three are now extinct and another thirty are currently endangered; corresponding figures among the forty-seven honeycreepers include fifteen extinct species and another sixteen endangered.

The palila, one of these endangered honeycreepers, occurs only on the big island of Hawaii. A striking bird, the palila possesses an unusually large bill, a golden yellow head and throat, and gray along its back. The palila became endangered mainly because of the destruction of its habitat resulting from agriculture and the introduction of nonnative species, principally goats and sheep. The grazing and browsing of these introduced herbivores destroyed vegetation necessary for the palila's survival, mainly the unique Mamani-naio forest. Most of this remaining forest remains confined to the upper slopes of Mauna Kea in Hawaii Volcanoes National Park.

The irony of the palila's plight stems from the historic perpetuation of these destructive mammals by the very government agency mandated to protect and manage the islands' wildlife, the Hawaii Department of Land and Natural Resources (HDLNR). Like other state wildlife agencies, the HDLNR depends on funds derived from the sale of hunting licenses and taxes on ammunition and sports equipment. At first, many viewed the introduction of land mammals as a positive biological contribution to the islands, given the absence of indigenous mammals other than species of bat, seal, and humpback whale. Like many isolated island environments, Hawaii had many unique organisms but relatively few overall number of species. Many of the early proponents of wildlife conservation, including hunters, birders, and wildlife officials, ironically advocated introducing foreign species. All appreciated how Hawaii's fair climate and lack of predators provided an especially hospitable setting for species introductions. The introduction of various goats, sheep, pigs, and other large mammals, both deliberate and accidental, further provided hunting opportunities for the islands' small population of sportsmen and revenues for its game agency.

Few realized at the time how devastating the impact of these invasive creatures would be to Hawaii's indigenous wildlife. By the time they did, most had become deeply committed to established practices and were unable to make the necessary changes to rectify the situation. Considerable concern developed among ornithologists and private environmental groups, however, regarding the damaging impact these exotic species were having on the islands' endemic birdlife, including the palila. Failing in their efforts to have the state remove the offending creatures, the Sierra Club initiated legal action in the early 1980s on behalf of the endangered honeycreeper in what became the famous case of *Palila v. Hawaii Department of Land and Natural Resources.* This legal action sought to remove the exotic herbivores from the Mamani-naio forest of Mauna Kea because of their allegedly destructive impact on the palila's habitat essential to its survival. Like the snail darter case, this conflict represented another classic confrontation of politically powerful and well-established interests versus emergent environmental constituencies and new wildlife values. This time, however, the spotlight focused on the very government agency responsible for managing and protecting the state's wildlife.

After many legal judgments and appeals, the courts eventually ruled in favor of the palila. The legal ruling also significantly expanded the commitment to protect endangered species in several important respects. First and perhaps foremost, the court determined that the destruction of the palila's habitat constituted a prohibited "taking" of the animal in violation of the Endangered Species Act. The idea of taking had until that time mainly focused on direct physical harm, removal, or killing of an endangered species. The palila judgment fundamentally altered this assumption by ruling that significant damage to the palila's habitat represented a taking no less harmful and prohibited than direct bodily injury or mortality.

A second critical aspect of the ruling against taking an endangered species was that it applied to any public or private land or property. Prior to the palila case, direct harm to an endangered species' habitat had been largely restricted to government-owned lands as in the snail darter case. The broadened definition of taking in the palila case vastly expanded the potential reach of the Endangered Species Act by including all public and private lands and waters of the United States. Any person or entity that knowingly engaged in destroying a listed species' habitat might be prosecuted for illegally taking an endangered species. This interpretation has ever since been the cause of considerable controversy. A recent court case attempted to repeal the palila judgment but failed to remove the destruction of habitat on private lands from the purview of the taking provision of the Endangered Species Act.

A third important consequence of the palila case stemmed from its focus on the destructive impact of introduced animals as a cause of species endangerment and extinction. As noted, introduced species have been responsible for endangering perhaps one-fifth of all vertebrate species, and their impact has been particularly devastating on islands and other geographically isolated environments. A major difficulty in dealing with this problem has been a lack of public awareness and understanding—in fact more than 98 percent of Americans and Japanese do not recognize this factor as a major reason for species endangerment and extinction. The palila case substantially expanded public and official awareness of this problem, especially in Hawaii.

A final important consequence of the palila case derived from its exposure of the distortions wrought by an unbalanced and inequitable public wildlife management emphasis on game species and consumptive wildlife use. The financial contributions of sportsmen have certainly been critical in the development of American wildlife conservation. But an unintended consequence of depending on taxing sportsmen to generate agency revenues has been a distorted stress on a small number of game animals, even destructive nonnative species. This game management focus has become less and less tenable in the current context of massive biodiversity loss and species endangerment, as well as American society's rapidly expanding interest in nonconsumptive wildlife use. The palila case dramatically illustrates the price of this misallocation of resources and distorted perspective.

Unlike the snail darter case, a remarkable degree of public support developed for the palila in Hawaii. This different political climate may have been a consequence of the limited economic stakes involved in the dispute. Another factor may have been differing public perceptions of an attractive bird species as opposed to an unknown fish. This notion—that people's attitude depends on the endangered species involved—is reflected in a national study result.[10] As Figure 22 reveals, most Americans support protecting popular and aesthetically appealing species like the bald eagle, mountain lion, trout, and American crocodile, even when this protection might result in significant increases to the cost of an energy development project. Only a minority, however, would accept such a sacrifice to conserve endangered species of snake, plant, or spider. These results reflect the significance of aesthetics, familiarity, and higher taxonomic

Figure 22. Attitudes Toward Protecting Endangered Species Despite Increased Cost of Energy Development Project

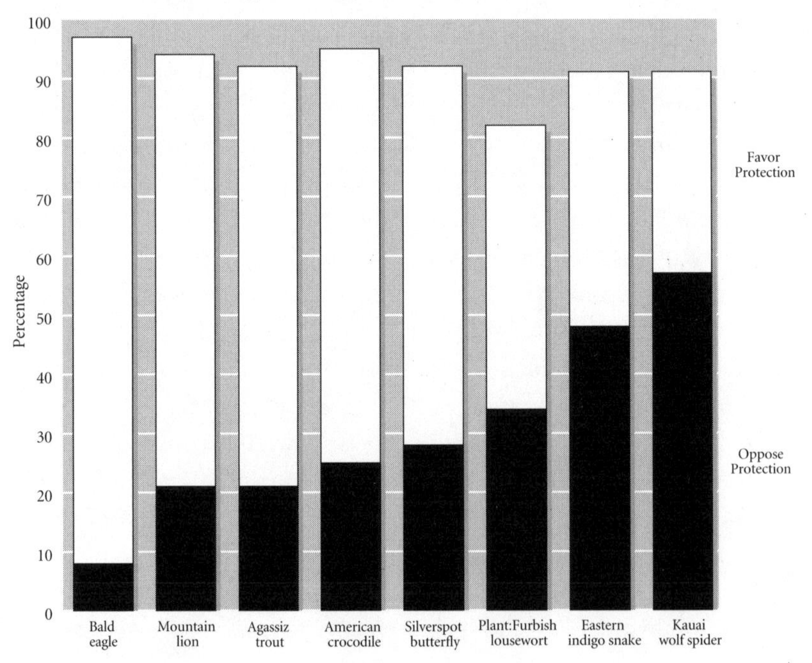

status in eliciting support for rare and endangered wildlife. The palila probably benefited from possessing many of these characteristics—as well as not conflicting with important economic development interests.

The Black-Footed Ferret

The final case study focuses on the story of the black-footed ferret, a victim of excessive persecution of its primary food source, the prairie dog, and massive destruction of its native ecosystem.[11] The ferret and prairie dog unfortunately stood in the way of settlement of the western prairie and were eventually pushed aside by a prevailing view of the land as mainly useful if rendered materially productive for humans. The black-footed ferret itself never achieved the status of despised or persecuted competitor. Indeed, many viewed the ferret as a cute little critter, although most hardly noticed the creature at all because of its tendency to remain largely underground and become active only at night. The demise of the black-footed ferret stemmed largely from pervasive antipathy and widespread persecution of the prairie dog, a diminutive and seemingly ubiquitous rodent of the western rangelands. The prairie dog was widely viewed as a competitor of cattle for forage, as well as a frequent cause of injury to livestock due to the innumerable holes of its "towns," sometimes stretching for tens and even hundreds of miles across the western grasslands. Antagonism toward prairie dogs eventually resulted in an eradication campaign perhaps rivaled only by the war against the wolf in its single-minded devotion to annihilating another form of life on the western prairie.

The black-footed ferret is a predatory member of the weasel family, the mustelids, its prey consisting almost exclusively of prairie dogs. Living largely within prairie dog communities, the ferret spends most of its life below ground seeking its prey mainly at night deep within the town's many tunnels. Indeed, the black-footed ferret was largely unknown even to the scientific community. Historical records, sampling, and prairie dog population estimates suggest that possibly one million ferrets may once have existed on the western plains.

Most ranchers view prairie dogs as the quintessential pest. These rodents, they believe, directly compete with cattle for grass, and their holes are a menace to livestock. Additional biases against prairie dogs include their attractiveness to undesirable wildlife such as coyotes and snakes and, like most rodents, the presumption that they cause disease. Most westerners thought distributing poison around prairie dog holes or routinely shooting the varmints constituted the mark of a good citizen. Black-footed ferrets may not have been especially disliked or well known, but they became a victim of guilt by association. Moreover, they appeared to possess little obvious value—even their pelt, unlike the fur of other mustelids such as the mink or fisher, was too coarse to market.

The war on prairie dogs eventually took its toll, even on a creature as widely distributed and prolific as this rodent. The animal became extirpated in one area after another; some prairie dog species even became endangered. The impact on the black-footed ferret was especially disastrous, given its dependence on large numbers of prairie dogs for predatory success. By the 1970s, this mustelid had presumably become extinct in the wild, as the last known population in South Dakota mysteriously succumbed to disease.

The ferret was miraculously "rediscovered," however, in the early 1980s near Meeteetse, Wyoming, within the shadow of Yellowstone National Park. This population of less than one hundred animals existed largely on private ranchlands, although some lived on public land as well. Despite close monitoring and study, disease struck again, leading to the population's rapid demise. Indeed, the species was almost extinct when less than a dozen survivors were removed in the late 1980s. A successful captive breeding effort followed, and a small number of ferrets were reintroduced into the wild near Meeteetse in the early 1990s. Not only has this restoration attempt been controversial, but the animal's susceptibility to disease underscores the continuing uncertainty regarding the creature's long-term prospects for survival. Future reintroduction efforts planned for South Dakota, Montana, and elsewhere could greatly enhance the ferret's chances for recovery.

The black-footed ferret's brush with extinction and the difficulties of its reintroduction illustrate the complicated multidimensional challenge of protecting and restoring endangered wildlife. The biological impediments have been many, particularly because this two-pound minklike animal has rarely been studied in the wild. Most of our knowledge derives from research on captive animals, and there is little reliable information on the ferret's feeding, reproductive, and other survival strategies in the wild. Current restoration efforts are frustrated by a remarkable degree of biological ignorance, uncertainty, and unpredictability.

Perhaps even more formidable have been the many socioeconomic, regulatory, and attitudinal obstacles to protecting and recovering this most endangered of North American mammals. A major socioeconomic hurdle continues to be the conflict over the use, control, and management of the public rangelands. Western livestock producers assumed long ago "de facto" ownership of much of the public grasslands for feeding and raising domestic animals.[12] These ranchers tend to resist all restrictions on their right to graze livestock and control pests on these lands. Their anxieties have been exacerbated by efforts to reform the management of the public grasslands, impose higher grazing fees, and restrict or even eliminate livestock activities that appear to threaten the survival of endangered species.

Ranchers and other groups who depend economically on extracting natural resources from the public lands and waters (loggers, miners, commercial fishers) often view themselves as possessing an inalienable right to exploit resources from these areas. They vigorously object to any threat to these presumed rights and frequently regard expanded regulation and associated legal actions as constituting a growing menace to their way of life and an attempt to seize control over public lands by outsiders, particularly government officials and environmentalists.[13] The current property rights debate represents the most recent in a long history of these disputes. Efforts to protect endangered species like the black-footed ferret often crystallize this perceived challenge to long-established prerogatives of access and control over public lands and resources. This polarization and conflict may be diminishing, however, as a new convergence among some ranchers and environmentalists has started to view restoration of biological diversity on the rangelands as in the long-run interests of wildlife, cattle, and humans alike. This still nascent development is discussed in greater detail in the next chapter.

Regulation is another obstacle to the black-footed ferret's recovery. The most serious problem has been the many laws developed to encourage cattle and sheep production, as well as to control wildlife competition and depredation of livestock. The result has been a complex of extensive relationships among ranchers, government agencies (particularly the Bureau of Land Management and the animal damage control divisions of the Departments of Interior and Agriculture), and western legislators. This "iron triangle" of political power has allowed stockmen to exercise considerable influence over the management and use of public lands.

Ranchers' influence has been particularly institutionalized in government agencies like the Bureau of Land Management and Department of Agriculture. These large bureaucracies have tended to act conservatively, following well-established patterns of behavior. As a consequence, such agencies often appear ill-equipped to make bold and innovative policy changes, especially in response to new, challenging, and unpopular regulatory mandates like the Endangered Species Act. Both tradition and training have directed these agencies to provide services that enhance commodity-production goals. Considerable political risk inevitably derives from limiting these traditional utilitarian services and the power of commodity interests for the sake of protecting rare and endangered wildlife. Long-established authority structures, coupled with a dominant culture of resource exploitation, have worked to maintain the status quo and thwart innovative environmental protection objectives. Laws like the Endangered Species Act and the emergence of new wildlife preservation constituencies appear to threaten this political and regulatory climate. For the black-footed ferret, the result has been limited agricultural support and considerable opposition to its reintroduction and recovery.

Attitudes have further contributed to the plight of the ferret. This factor has been especially cited by Richard Reading in his study of Montana ranchers, environmentalists, and the general public regarding proposed black-footed ferret reintroduction in this state.[14] Reading revealed that most ranchers, particularly those living near the proposed reintroduction site, hold very antagonistic views toward the possible restoration of this endangered species, especially the idea of protecting prairie dogs to assist in ferret recovery. Most livestock producers, rather than expressing actual dislike of the creature, believe black-footed ferrets have little practical or ecological value. Prairie dogs, however, incense ranchers and are generally viewed as destructive and worthless. As one rancher exclaimed: "I'm not against the ferret. It's the prairie dog. I want to kill every last one."

The gravest threat posed by the reintroduction of the black-footed ferret to most ranchers is the feared loss of control over the public rangelands. Many ranchers possess an almost xenophobic hostility toward outsiders, often viewed as interfering urban and foreign elements. Proposals to protect black-footed ferrets, prairie dogs, and endangered species are regarded by many as a fundamental threat to traditional powers, property rights, and the rural ranching culture. Many fear the land and its resources will be "locked up" and historic values of freedom and exploitation of natural resources replaced by increasing government control and regulation. Many also believe that traditional consumptive uses like hunting and trapping will be discouraged in favor of nonconsumptive wildlife recreation. As another rancher remarked: "Ferrets give the environmentalists a way to get in and try to take over. . . . I don't mind the ferrets coming back, but it's all the restrictions that come with it." Many ranchers and rural residents, therefore, oppose black-footed ferret reintroduction because they believe it will bring major new constraints on economic development, the closing of roads, and designation of additional wilderness areas. Few stockmen think they will actually be precluded from protecting their livestock and crops from wildlife damage, but many fear they will be prevented from reducing prairie dog populations, especially using poisons.

Reading's study explored basic attitudes toward black-footed ferrets and prairie dogs among ranchers, environmentalists, and the Montana general public. Montana ranchers' attitudes toward ferrets and prairie dogs were remarkably similar to farmers' views in Minnesota and Michigan toward wolves. Montana ranchers voiced strong utilitarian and negativistic attitudes and consistently opposed government control or limitations on resource use to protect or conserve ferrets and prairie dogs. Most regarded ferrets and especially prairie dogs as having little emotional appeal or naturalistic value, and they strenuously objected to the notion that these animals possess moral rights. Moreover, these unsympathetic attitudes had little to do with knowledge of these animals. In fact, ranchers were among the most informed of all groups about prairie dogs and black-footed ferrets.

Urban Montanans and especially environmentalists regarded these creatures quite differently. Generally they viewed black-footed ferrets and prairie dogs with considerable affection, even appreciation, and strongly supported the ferret's restoration and recovery. Even in a state as homogeneous as Montana, urban residents and environmental group members contrasted greatly with ranchers and most rural residents.

Overall support, nonetheless, existed for saving the black-footed ferret from extinction. This animal undoubtedly has benefited from its aesthetic appeal—it has a round face and a patch over its eyes—and from being labeled the most endangered mammal in North America. The creature's "rediscovery" in the shadow of the continent's most famous wilderness area has also contributed to its support. Even so, its invariable association with prairie dogs, and threats to agricultural control over the use of public lands, constitute major obstacles to the ferret's restoration. An array of biophysical, socioeconomic, institutional-regulatory, and valuational factors continue to plague the black-footed ferret's chances for recovering from the abyss of extinction.

A Few Lessons

What can be learned from the collective destinies of the snail darter, palila, and black-footed ferret about protecting and recovering endangered species? This question defies easy answer and its full consideration may be beyond the scope of this book. Nevertheless, some suggestions can be offered.[15]

The goal of endangered species protection and recovery clearly depends on an adequate understanding of the basic values, motivations, and interests of all stakeholders. Most endangered species programs, however, fail to consider the wide diversity of values held by competing constituencies in an endangered species dispute. The great majority of recovery programs rarely ascertain prevailing attitudes or recognize how beliefs can profoundly influence the prospects of a species' survival. Values tend to be viewed by management professionals as vague and insubstantial, undeserving of scientific attention, or frequently assessed in intuitive and biased ways. A consistent problem in many endangered species programs is an underestimation of key value differences among critical stakeholders and an inadequate incorporation of this information in the design and implementation of recovery efforts. All wildlife values, however, must be fairly and systematically considered if the goal of endangered species protection has any chance of advancing.

A reluctance to assess values scientifically often stems from a concern about producing quantitative evaluations of what seems unquantifiable. This challenge reflects the difficulty of assigning numerical equivalents to such endangered species values as the mythic qualities of a grizzly bear, or the prowess of a rare fish in the rushing surge of a snowmelt spring, or the mysterious cries of a whooping crane calling across what sounds like the vast sweep of time. These values obviously remain elusive and nearly impossible to capture, yet the snail darter case suggests that omitting these qualities of living diversity can engender, almost by default, decision making inherently biased toward the economically quantifiable. Numeration and empiricism may be a crude and blunt instrument, but modern society continues to be governed both in custom and law by a tyranny of numbers. If all wildlife values fail to be systematically assessed and measured, policymaking will almost inevitably be weighted toward commodity production and marketplace objectives. A more rational approach suggests that all values of living diversity should be scientifically and equitably considered among the widest range of stakeholders involved in an endangered species dispute.

Considering the views of all constituencies does not mean all perspectives are equally valid—especially when certain views contribute to the extinction of a species. Effective biodiversity conservation will also necessitate altering prevailing attitudes by providing information that will help people make intelligent and informed decisions. Beyond the fundamental importance of science, policy, and management, education and ethics represent critical tools of endangered species conservation. The destiny of most creatures depends on human knowledge, values, and beliefs. When people recognize that their chances for a richer and more rewarding life hinge on a healthy and diverse biota, they will be more inclined to make choices consistent with the preservation of endangered species.

The case of the black-footed ferret, as well as the wolf and whale, clearly suggest that expanded appreciation and concern for endangered species rarely derive from simply increasing people's knowledge. A common fallacy of many endangered species programs has been the naive assumption that greater factual understanding will encourage greater public support for species restoration and recovery. Yet basic values tend to be more deeply entrenched and resistant to change. Most people use additional information simply to rationalize and reinforce their beliefs rather than to alter them. Ardent opponents and proponents of black-footed ferret, wolf, and whale recovery often possess similarly high levels of knowledge despite their widely divergent values. Education about endangered species necessitates addressing people's basic attitudes and motives as much as their factual understanding. Support for endangered species conservation will emerge when people believe this effort enhances the prospects of a more materially, emotionally, and spiritually worthwhile life for themselves, their families, and communities. This may not constitute a particularly easy task, but it may be an unavoidable one.

The experience of the snail darter, palila, and black-footed ferret further emphasizes the importance of considering various socioeconomic and political forces in endangered species protection and recovery programs. These factors stress the significance of many power, property, and authority relationships within and among communities and societies. Local groups tend to resist change, often perceiving endangered species protection by government agencies and environmentalists as insensitive and hostile to established relationships and customs. Rural, resource-dependent groups frequently oppose endangered species programs as much because of the challenge to their traditional prerogatives as because of particular hostility to an imperiled species.

Most endangered species programs will fail, however, if they lack local support. A commonly neglected aspect of this problem is the issue of property rights—particularly threats to traditional uses of private lands and access to resources on public lands. Endangered wildlife often exacerbate these property rights issues by moving across jurisdictional boundaries, including public and private lands. In the black-footed ferret program, local communities often expressed anxiety about the threat posed to historic grazing rights on public lands, as well as their ability to control prairie dogs on private and public lands. Traditional rights associated with private landownership and public land access have become poignant symbols in this country, particularly in the western states. Effective protection of endangered species requires respecting these rights, as well as demonstrating that the productivity of the land ultimately depends on the ecological health and diversity of all its biota.

Community involvement in formulating and designing endangered species programs could well increase the motivation to support these efforts. A frequent impediment to local participation is the failure to identify the tangible benefits to communities that accrue from embracing the virtues of endangered species protection. Some inducements for strengthening trust and involvement might include expanding local decision-making authority, tax incentives, agricultural subsidies, and citing the economic returns from biodiversity conservation (ecotourism, new products, more reliable water supplies, increased agricultural production).

Another property rights problem of many endangered species programs stems from the diversity of local, state, and federal authorities, each with somewhat different responsibilities and stakes in species protection and resource use. Intergovernmental competition and confusion can lead to conflicts among government agencies, especially state/federal relations, regarding the control of endangered species. The states have traditionally managed resident wildlife, while federal authorities have assumed primary responsibility for migratory animals. The passage of the Endangered Species Act represented a major expansion of federal powers. The substantial growth of powerful national environmental organizations has also encouraged a view among states and local communities that both the "Feds" and "Tree Huggers" are using the plight of imperiled wildlife to expand their control over public and private lands and resources.

It will not be an easy task to defuse these state/federal and national/local tensions. Increasing endangered species management responsibilities at the state and local level could help, especially if federal authorities and private environmental organizations function more in the role of providing expertise and partnership assistance. Devolution of authority to the local level can diminish hostility toward government agencies and outsiders often perceived as having little stake in traditional economies, lifestyles, and institutions. Nonetheless, empowering local entities runs the risk of parochial elites and special interests assuming inordinate influence and pursuing objectives inconsistent with endangered species protection. Comanagement efforts involving local stakeholders and government agencies could address community interests within a context of endangered species conservation. Cooperation and partnership can benefit from the accumulated wisdom of local communities, while maintaining the important role of scientific expertise and emphasizing the government's broad public interest.

Organizational factors should be considered as well in pursuing endangered species protection objectives, particularly the behavior and structure of government agencies. The subject of organizational behavior is one of the least understood and appreciated elements of endangered species program success or failure.[16] The snail darter, palila, and black-footed ferret cases clearly demonstrate that the behavior and commitment of government agencies represent a major factor in the fate of many imperiled species. The challenge of making government agencies perform effectively and efficiently constitutes a difficult goal indeed. As one policy analyst remarked: "Whether policies succeed or fail . . . is largely dependent upon the nature of the organizations mandated to carry out these policies."[17] The survival chances of many species will hinge on the capabilities of various government agencies to protect and conserve these creatures.

A number of internal and external organizational factors have been cited as the keys to effective government performance. All must be considered in designing successful endangered species programs.[18] Key internal factors include: an agency's basic goals and objectives, its procedures for measuring behavior and performance, incentive and reward structures, leadership and authority patterns, culture and ideology, communication and information flows, professional specialization, and division of labor. Key external factors influencing an agency's behavior include: its sources of funding, traditional constituencies and adversaries, controlling politicians and officials, public perception, and media treatment. Together these internal and external factors constitute an array of forces substantially influencing the capacity of wildlife agencies to protect and recover endangered species. Indeed, these factors determine the destiny of imperiled creatures just as much as understanding biology, human values, or critical socioeconomic and political structures.

Government agencies, however, often lack sufficient incentive to examine and ameliorate their established organizational structures and standard operating procedures. Bureaucratic interests frequently reside in maintaining a low profile and avoiding risky actions that might provoke powerful political and economic interests. This tendency toward the status quo is often characteristic of government bureaucracies operating in a noncompetitive context and lacking clear performance standards. Such agencies frequently avoid controversy, innovation, and risk—behavior often essential in endangered species programs attempting to alter years of established customs, traditions, and institutions.

Most endangered species programs tend to be timid. They are inclined to measure performance more by the maintenance of standard operating procedures than by explicit performance criteria such as the number of species listed or delisted, harmful actions prevented, amount of habitat protected, or recovery plans developed and implemented. Endangered species conservation requires agencies to be flexible and problem-oriented. Recovery programs tied to rigid operating procedures and authority structures tend to be less successful than adaptive and consensual decision-making efforts. Organizations, like people, need to learn from their mistakes, encourage corrective feedback, and work toward achieving an ongoing capacity for improvement and renewal.

Even with idealism, knowledge, adequate resources, widespread public support, and effective leadership, successful endangered species programs will be exceedingly difficult to achieve. Yet, the enormity of the contemporary extinction crisis more than justifies the effort. Wildlife agencies must become more effective, more efficient, and more equitable in managing and protecting endangered species. But for this to happen, these agencies must aspire to become the biological arm of government, unabashedly advocating the cause of species conservation and eschewing the more modest role of simply providing services to traditional interests and constituencies.

Conclusion

Modern society has never seemed more inclined to recognize the importance and legitimacy of saving endangered wildlife. We seem to be making slow, incremental progress toward altering the inertial flow of a historic process that for centuries has devastated much of this country's and the world's living diversity. Every small measure of success, however, seems to be countered by another major loss of habitat and species eliminated as our technological reach, per capita consumption, and explosion in human numbers continue their seemingly inexorable path of destruction.

Protecting and restoring endangered wildlife will require considerable scientific and political expertise. Recovery programs will need to include a far more thorough understanding of diverse biophysical, socioeconomic, valuational, institutional, and regulatory factors. The role and impact of competing stakeholders will also need to be examined and comprehended. The success of endangered species programs will ultimately depend on decisions made by ordinary people pursuing their daily lives. Imperiled wildlife will continue to cling precariously to survival until their destinies are positively linked to the social, political, economic, and psychological experience of most people and their communities. Policymakers must recognize that the complexity of endangered species conservation will necessitate multidimensional knowledge and interdisciplinary approaches.

A remarkable degree of ignorance, uncertainty, and unpredictability must also be acknowledged. Deficient understanding will not be easily remedied, and endangered species programs must proceed with caution and humility. Inadequate knowledge should not be an excuse for inaction; nor should it be an invitation to act with arrogance as if we know all we need to know. The prudent path will be one of restrained decisiveness, building buffers against the inevitability of surprise and unknowns in managing imperiled wildlife. Recovery programs will need to proceed with an action tempered by modesty, recognizing the limits of existing knowledge, the potential for disastrous mistakes, and the awesome implications of extinction.

Saving endangered species faces the intrinsic difficulty of being both reactive and crisis-oriented. Preventing species extinction represents the biological equivalent of emergency room admission: there is little margin for error and often compromise. The best alternative may be to implement protection *before* species confront the imminent prospect of extinction, strategically emphasizing the conservation of species and habitats of particular importance and a reasonable potential for recovery. Assigning priorities to endangered species protection may be ethically hard and biologically problematic, but reasonable criteria exist for identifying creatures of special significance. Biologically, as Reed Noss suggests, this might include "keystone" species whose existence determines the fate of other dependent plants and animals, "indicator" species whose condition represents the health of particular habitats and ecosystems, and unique species possessing genetically rare and important biological characteristics.[19] Sociological priorities can be assigned, as well, such as "flagship" species of particular historic, aesthetic, or cultural significance, whose protection may engender public support and thus educate others about the importance of a healthy and diverse biota. A final priority could include "umbrella" species occupying such a large territory that their conservation protects various other creatures.

These difficult choices underscore the indispensability of expanded public understanding and appreciation of living diversity, above all its rare and endangered forms. Conservationists must strive to convey to the public and its policymakers the grievous burden of imposing the loss of so much biological capital on future generations. All species represent an extraordinary array of biochemical and physical adaptations hammered into their genes after countless generations of evolutionary trial and error. It would be tragic to forsake the practical benefits that might be derived from extinct species just when society's knowledge of molecular biology and bioengineering intimates a revolutionary increase in the capacity to exploit a species' unique and irreplaceable biophysical characteristics. Beyond these material rationalizations, we need to describe how human lives filled with intellectual promise and spiritual meaning depend on a rich and diverse biota. The diversity of life remains at the core of human growth and development, and the massive eradication of other creatures invariably degrades our own mental and ethical as well as physical foundations. Few parents knowingly choose to bequeath to their children a world more impoverished and degraded than the one in which they themselves struggled to find meaning and purpose. People will opt for maintaining and restoring living diversity when they recognize its indispensable role not only in their own lives but in the lives of their future descendants as well. The challenge of endangered species conservation invites us to educate others, and ourselves, about the many values that inextricably link the fullness of human experience to what Joseph Wood Krutch called the great chain of being.[20]

Fundamental reform of our wildlife institutions appears to be a necessity if we are to design programs capable of confronting the contemporary extinction and biodiversity crisis. We may no longer have the luxury of waiting for the promise of incremental change. If current institutions rise to the challenge of today's global loss of species, future generations will laud them for their courage, foresight, and wisdom. If they fail, they may be considered as guilty of contributing to a dark age of biological apocalypse as the more obvious reasons for endangerment and extinction.

CHAPTER 8

Conserving Biological Diversity

THIS CHAPTER considers various issues involving the conservation of biological diversity—especially the role of values in the pursuit of this elusive objective. Our focus here is on human competition with, and exploitative use of, biological resources. As competitive exclusion of other life forms can occur in many ways and in various contexts, far more than might be reasonably covered in this chapter, this discussion illustrates human impacts on biodiversity by examining two kinds of rural land uses, agriculture and forestry, and by considering issues of biological conservation in the modern city. These explorations emphasize the importance of engendering a more positive and reinforcing relationship between people and biological diversity as a basis for achieving lives of prosperity, value, and satisfaction. We then turn to the issue of human consumptive and nonconsumptive uses of biological resources—especially the commercial trade in wildlife. The chapter concludes with some reflections on basic changes in wildlife management needed to meet the contemporary challenge of human competitive exclusion and overutilization of biological resources.

Competition

The global loss of biological diversity is largely a consequence of increasing competition, both direct and indirect, between people and wildlife. Direct competition between people and other life forms for the same resources was one of the issues at play in our discussions of controlling wolf and prairie dog impacts on livestock. Direct competition has historically inflicted serious harm on many species—particularly when there was less ambivalence regarding the legitimacy of eliminating wildlife that conflicted with human wants. Direct competition, at least with vertebrate species, has diminished greatly in modern times. Indirect impacts on wildlife habitats and ecosystems, however, have expanded exponentially as a consequence of huge increases in human numbers, technology, and per capita consumption of energy, space, and materials. Human co-optation of wildlife habitat represents today by far the most serious, pervasive, and insidious threat to biological diversity. People have captured and usurped so much of the earth's available energy and space that massive reductions in the abundance and diversity of other life forms have occurred.[1] This loss has resulted not just from the decline in total available wildlife habitat, but also from the impact of fragmented and simplified landscapes.

FORESTRY AND AGRICULTURE

Modern agriculture and forestry especially illustrate the effects of human competitive exclusion of wildlife.[2] Today most large-scale agricultural and forestry operations have encouraged the massive simplification and homogenization of vast stretches of landscape. The single-minded pursuit of commodity production goals, as well as the destructive technology employed, have wreaked tremendous damage on many creatures and habitats. Large-scale crop and timber production, theoretically at least, does not have to result in the destruction and exclusion of so much nonhuman life, but the operational reality tends to be devastating in its impact. Commenting on much modern industrial forestry, Starker Leopold observed:

> In my own experience, ground level decisions . . . almost universally favor timber production on fair to good sites. Roads are built, timber is cut, pines are planted, snags are felled, brush fields are bulldozed, herbicides and pesticides are applied—with inadequate regard for ecosystem effects. Many or all of these . . . procedures are locally justified to advance the objective of growing timber for higher production. My criticism concerns the *totality* of devotion to production goals which seems so often to take precedence over other publicly declared objectives of . . . land management.[3]

Many damaging forestry and agricultural practices stem from an overwhelming emphasis on producing the maximum number of just a few commercially favored species cultivated under intensive and highly artificial conditions. This scale of single-species or limited-species production demands the suppression of other presumably competing organisms—frequently by widespread and indiscriminate application of varying toxicants and biocides. Biological impoverishment is further accelerated by transforming many natural areas, such as wetlands and slopes, into uniform habitats more amenable to large-scale production and machinery. Industrial forestry operations often contribute to ecological simplification by eliminating various forest growth stages, particularly older-age and climax phases of mature forests, as well as retarding plant decomposition and nutrient cycling by removing forest litter, downed trees, snags, and understory vegetation. Extensive road building can further lead to soil loss and compaction, while agricultural planting, fertilizing, spraying, and harvesting over enormous stretches of homogenized landscape frequently degrade and impoverish landscapes and ecosystems.

These and other destructive practices of modern agriculture and forestry have been widely documented and described. Often the result is a massive loss of biological diversity and impairment of many critical ecological processes. The key question, of course, is: can anything be done to reverse this degree of genetic decline and impaired ecosystem functioning given the imperatives of modern economics and the requirements of a highly populated and increasingly urban society? Innovative environmental laws and public health standards have certainly mitigated some of the more harmful practices, particularly the indiscriminate use of certain biocides and destructive harvesting operations. New technologies and new biological knowledge also indicate the potential for developing less injurious forms of agriculture and forestry.

This book, however, is not about new technologies, laws, and biological knowledge. It is about the role of values in determining the kind of world in which farmers, foresters, and ordinary citizens choose to live and sustain themselves. Basic perceptual shifts will be as important as the newest technological innovation, scientific discovery, or regulatory mandate in resolving the current biodiversity crisis. Fortunately, various signs suggest that significant and salutary value changes may be occurring among an important segment of progressive farmers and foresters.

Many have begun to question the long-term lifestyle and even commodity production impacts resulting from destroying so much genetic variability and natural process. Both personal experience and increased understanding have suggested to many foresters and farmers that adequate levels of agricultural and timber production can be sustained over the long run only in a context of healthy ecosystem functioning and an associated array of species interactions and interdependencies. More and more foresters and agriculturists recognize that soil fertility, clean and abundant water supplies, the control of pests, plant pollination and reproduction, forage availability, and various other critical elements of adequate commodity production depend in many known and unknown ways on an abundant, diverse, and healthy biota. Moreover, many landowners have concluded that their long-term options for generating economic returns from the land depend on the capacity to adapt and respond to evolving opportunities associated with the discovery of new practical uses of indigenous plant and animal species.

Many agriculturists and foresters have also come to recognize that healthy, diverse, and attractive natural systems represent an irreplaceable ingredient in individual, family, and community quality of life. Many have begun to question whether the industrialization of modern agriculture and forestry has alienated people from the land and from one another. Many have come to look more skeptically upon the pervasive unattractiveness and health-threatening aspects of many monocultural forestry and agricultural operations. Ironically, these rural landscapes were once celebrated for their amenity values, although today they appear to be increasingly viewed as deficient in quality and even dangerous. More and more farmers and foresters are realizing that all the relevant social and biological costs of large-scale industrial production must be assessed in making economic and lifestyle decisions over the long term. Based on these new calculations, many have begun to question whether they are still pursuing worthy and satisfying lives marked by pride and pleasure—or have embraced instead an existence increasingly characterized by ugliness and risk for their families, communities, and future descendants.

An increasing number of farmers and foresters are thus concluding that a key ingredient in a more nurturing lifestyle and sustainable economy is a view of the natural environment as a community first and a commodity second. They believe this will necessitate learning to love the land again and the values of its many creatures. They recognize, too, that in restoring health and honor to the landscape they can reconnect with themselves and one another, as well as enrich the earth's potential for producing more physically, emotionally, intellectually, and spiritually satisfying lives. Healthy and attractive landscapes are increasingly viewed as providing people with cultural meaning and communities marked by familiarity and security. People still utilize and exploit the land as they must, but with a care and gentle prudence that respect the biological enterprise as an ancient and proven economy and social order.

The erosion of the connections between people and the natural landscape represents as much a challenge to modern agriculture and forestry as the mitigation of public health threats posed by the seepage of biocides into food chains or declining commodity production caused by damaged ecological functioning and habitat destruction. As Alan Grussow has suggested: "It is not simply nostalgia for a romantic and rural past that causes us to grieve over the loss of natural open space. It is a concern over the loss of human values. For we are not distinct from nature; we are a part of it, and so far as our places are degraded, we too will be degraded."[4]

THE MODERN CITY

The logic applied to agriculture and forestry can also be extended to the modern city.[5] One of the tragic assumptions of contemporary urban life is a belief that city people no longer need an abundant, diverse, and healthy biota to lead lives rich in satisfaction and meaning. Urban open spaces have consequently been treated with callous disregard as if their value derived only from their private or municipal development. Few resources are directed at maintaining or enhancing the various aesthetic, naturalistic, ecologistic, or even utilitarian values of most urban natural areas. Most city dwellers are repeatedly reminded of the presumed unimportance of biodiversity in the urban context. Environmental values are routinely omitted from considerations of architectural design, siting decisions, road placement, industrial development, shopping center construction, and various other land management and planning choices. The average city official views with indifference the remaining pockets of open space, often seeing expenditures on the environment as the lowest of budgetary priorities. Rare is the urban developer and politician who recognizes the links between a city's long-term economic viability, its healthy functioning, and the quality of its natural landscapes.

These assumptions are both unfortunate and ultimately self-defeating. Natural diversity may be less dramatically evident or conveniently available in the modern city, but it nonetheless remains essential to the long-term emotional and material well-being of urban dwellers. City people still depend on ongoing opportunities for intimate and rewarding experience of natural diversity to achieve full and satisfying lives.

Ironically, most of the great cities of the world were originally sited because of the abundance and diversity of their natural values. The majority of American cities, for example, were located along coasts, shorelines, lakes, estuaries, rivers, mountains, and other prominent natural areas distinguished for their environmental appeal and biological richness. These attributes attracted people to settle because they offered compelling opportunities for personal, family, and community development. These areas struck most people as laden with the potential for sinking deep emotional and intellectual roots and establishing an enduring sense of place and connection. Most modern cities have lost sight of these initially compelling environmental virtues and attractions, frequently choosing to deny their citizens the chance to experience intimate and satisfying contacts with the natural world.

My home city of New Haven offers an instructive illustration of the prevailing bias. The city has, for example, a major highway along its remarkable harbor constituting a nearly impenetrable barrier separating people from a waterfront historically distinguished by its beauty and natural diversity (an area still known for its abundance of migrating shorebirds, wintering ducks, and shellfish). The nearly 20 percent of New Haven remaining in parks and open space are typically treated with indifference, as a safety hazard, or as a potential site for development. Few municipal leaders view these areas as opportunities for citizen renewal, outdoor recreation, or a chance to reconnect and nurture a host of environmental values. The natural amenities of these parks and open spaces remain largely unknown inside and outside the city, rarely trumpeted by comparison to the city's more publicized crime rate and numbers of homeless people and drug addicts. The prevailing assumption, as in so many other American cities, is that the promise of socioeconomic resuscitation lies in industrial development, more construction, and expanded roads and highways. New Haven, like so many other cities, has achieved little success in attracting or retaining businesses, despite extensive tax incentives, subsidies, and infrastructural support. Material and monetary bribes alone have failed to render these environmentally degraded cities more appealing or marketable.

Most urban leaders and developers have neglected to recognize that businesses, like individuals and families, choose to locate in areas for reasons beyond simply financial and logistic considerations. People and companies alike often make choices based on the quality and attractiveness of their environments: they want good places to live, rear, grow, and develop. Many if not most of America's most vital cities—places like Boston, Denver, Salt Lake, San Francisco, Seattle, Tucson, and Boise—represent areas still characterized by considerable natural beauty and environmental amenities. People and business seek these healthy and appealing landscapes, and their economic decision making is frequently driven as much by concerns for noneconomic environmental values as by a narrow cost-benefit calculus. Businesses follow people as often as the reverse—particularly when people, for a variety of environmental reasons, appear to be more contented and motivated. Good economics frequently equates with good ecology. Cities like New Haven would be well advised for commercial objectives alone to consider more seriously the environmental characteristics that have often made the suburbs and other areas more alluring to people and companies alike.

Many people believe insufficient opportunity exists in most cities for meaningful contact with the natural environment. This perspective reflects, however, a false dichotomy often made between cities and nature. Biological diversity occurs in surprising abundance in any city, and the potential for a rich and rewarding experience of nature exists in all but the most impoverished urban landscapes. As Edward Wilson has suggested, even a handful of urban soil contains more biological richness and complexity than all the dead planets of the solar system put together. Aldo Leopold similarly observed that the lessons of ecology can be found just as much in a city lot as in a redwood grove.[6] The meaningful experience of nature in the city constitutes more a challenge of design than an intrinsic flaw of modern urban life—if planners and developers are willing to scratch the surface of their own values as well as the landscape around them.

Satisfying and sustainable connections between urban residents and the natural world will require leaders, politicians, architects, businesspeople, and civic officials with the vision and courage to incorporate biotic diversity into the urban landscape. Great opportunities and even economic rewards await those planners, developers, and entrepreneurs imaginative enough to capture the aesthetic and ecological virtues of the natural environment and weave them into the lives of urban families, neighborhoods, and the places where people work. People even crave contact with nature in the modern skyscraper. This point is illustrated by the finding that workers in these environments are more likely to struggle, through posters, pictures, and potted plants, to reinsert living diversity into their often lifeless workplaces.[7] This vicarious effort typically proves insufficient, however, leaving many workers with vague feelings of frustration and perhaps diminished productivity.

The health and vitality of city people depends on affiliating with nature and living diversity in aesthetically attractive, ecologically sound, and materially accessible ways. The prevailing urban malaise of air and water pollution, habitat destruction, and denaturalized environments represents neither a necessary nor an inevitable reality. Litter, waste, and biological impoverishment constitute both ugly and debilitating conditions that can and ought to be replaced by natural grace and charm.

Many city dwellers remain disconnected from their natural surroundings as sources of meaningful recreation and sustenance. This dissociation from nature constitutes a serious limitation of many modern urban areas as places of residence and community. Cities will elicit their greatest loyalty, commitment, and consequent stability when they function as sites where people can confidently and consistently encounter satisfying connections with natural as well as economic and cultural wealth. If people experience cities as environmental outsiders and transients, they will likely behave toward these places with indifference and irresponsibility. The decline of the modern city may be due as much to the degradation of its natural amenities as to the more generally acknowledged economic and social reasons. Humans need continuous and spontaneous affiliations with the biological world, and meaningful access to natural settings is as vital to the urban dweller as to any other.

Many people believe the modern city can ill afford the luxury of maintaining or restoring healthy, diverse, and attractive natural environments. This may constitute, however, a narrow and shortsighted calculation. Individual meaning, community pride, and even economic survival very likely depend on attractive, rewarding, and fulfilling contacts with nature. Cities paralyzed by despair over the degraded state of their natural environment or by the presumed cost of its remediation may in fact impede their revival. Many cities have, fortunately, begun to marshal the will and resources to restore their natural values through community gardens, urban forestry and wildlife programs, wetlands renewal, greenways, park development, and more. These represent commendable efforts, but just the tip of the iceberg of what can be done. Cities will need to integrate nature and living diversity more broadly and more meaningfully into the pulse of urban neighborhoods, workplaces, and leisure settings.

Utilization

Human destruction and degradation of natural habitats represent the gravest threat to living diversity. Direct exploitation of wildlife, however, remains a major factor in the decline of many species. Excessive and unsustainable utilization of wild living resources continues to be a significant contributor to the endangerment of nearly one-third of all vertebrate species. Contemporary affluence, human overpopulation, and the scale of modern technology and transportation have meant that few species of significant economic value can withstand the unregulated impact of their commercial trade.

Despite the dangers of overexploitation, wildlife constitutes a renewable resource. Its utilization is neither intrinsically harmful nor destructive. The management challenge is how to achieve a sustainable balance between wildlife utilization, a species reproductive capacity, and its ecological contribution, all the while recognizing how much biological ignorance, uncertainty, and unpredictability exists. Another challenge is how to exploit these species while ensuring ethical standards of kindness and compassion. In a world of global markets and technology, sustainable wildlife consumption can only be achieved if attitudes of restraint, appreciation, and respect prevail for the bounty nature provides.

Wildlife utilization includes both consumptive and nonconsumptive uses. This distinction often masks the potential of any wildlife use to damage living diversity and natural habitats. All forms of wildlife exploitation can inflict significant and irreversible effects on species, biological communities, and natural habitats. For many people, nonconsumptive wildlife use may imply benign utilization because it typically lacks the deliberate intent to consume or destroy a life in the wild. Nonconsumptive use, however, can result in excessive and harmful consumption of wildlife habitat physically, spatially, behaviorally, and psychologically.8 Intensive nonconsumptive wildlife recreation in an era of high technology and mass transport can inflict cumulative impacts on species and ecosystems as injurious as any consumptive use. Nonconsumptive use can produce overcrowding, excessive energy use, air and water pollution, extensive hotel and highway construction, and other impacts that seriously threaten the existence of other life forms. The idea of benign nonconsumptive use often constitutes a myth with dangerous consequences—especially when it suggests there are no ethical obligations to behave with care, restraint, and compassion toward the living world.

All wildlife users carry the potential to leave the dark spoor of human excess on creatures and their landscapes whether through ignorance, indifference, or malice. Wildlife exploitation represents a privilege, not a right, with associated economic, ecological, and ethical duties. When humans deliberately enter into the world of another creature, whether for commerce or amusement, an obligation results. This utilitarian transaction obliges people to behave with skill, understanding, respect, and a sense of wonder and appreciation for the diversity of life.

Wildlife utilization is distinguished, too, by whether its primary purpose is commercial or noncommercial. Every form of wildlife exploitation inevitably has economic consequences, but commercial exploitation focuses foremost on a species' potential for generating profits and monetary returns. Sustainable commercial wildlife use represents a complex challenge. To illustrate the complexity of this topic, consider the international wildlife trade.

The international trade in wildlife has been highly contentious with a long and sad history of excessive exploitation resulting in the decline and extirpation of many species, exemplified earlier in the book by the great whales.[9] Historically, commercial wildlife exploitation has tended to maximize short-term profits even when it results in economic if not biological extinction, particularly when large consumer demand exists alongside a technical capacity to harvest and distribute products widely. Despite the illogic of eliminating the very source of the trade, people have frequently demonstrated a willingness to disregard future economic or biological consequences—particularly if there is significant opportunity for reinvesting surplus profits in other arenas of financial return.

Commercial overexploitation of wildlife is quite common in situations involving valuable species with a slow reproductive biology, especially when there are other species with physical characteristics that can be substituted for the commercially targeted animal. The rhetoric of sustainable wildlife use often implies strong incentives to maintain a steady stream of monetary returns in perpetuity, but the historical reality has frequently been otherwise. Harvesters have often behaved as if valuable species were inexhaustible, substitutable, or future technologies would mitigate or render irrelevant the exhaustion of the resource. People often discount the future, assuming that current economic returns and the power of human ingenuity will more than compensate for the risk and uncertainty of a species' elimination. These assumptions are exacerbated by the widespread prevalence of human arrogance, ignorance, and an unwillingness to enforce mechanisms for assuring sustainable management and species protection.

The international wildlife trade today yields some $5 billion to $10 billion annually. Despite existing regulations, many creatures continue to be victimized and endangered by the trade including various species of mammal (rhinoceros, tiger), birds (certain parrots, falcons), reptiles (species of crocodile, sea turtles), fish (species of tuna, ornamental fish), even invertebrates (some butterflies, shells) and plants (particular orchids, cactus). The extent of the current economic return and the hypothetical promise of sustainability have prompted many to promote the virtues of the trade as a viable economic development and conservation strategy, particularly for many developing countries.

The African elephant has been among the most controversial species associated with the international wildlife trade. A review of arguments for and against the marketing of elephant ivory highlights the many complexities and difficulties of commercial wildlife use and its effective implementation.[10] The elephant has become a focus of controversy largely because of two major characteristics of its complicated relationship with humans and the natural world.[11]

First, the elephant fits all the categories noted previously as conferring priority status on the conservation of a species: genetic, keystone, indicator, umbrella, and flagship significance. The genetic uniqueness of elephants derives from this animal being the only member of its order, Proboscidea, which includes just two species, the African and Asian elephant. By comparison, the mammalian order Carnivora comprises all the cat, dog, bear, mongoose, weasel, and hyena species. The African elephant's "keystone" importance, particularly the bush elephant of East Africa, is reflected in its dramatic impact on the condition and character of the East African savanna. The size, power, and physical adaptability of elephants have enabled these creatures to control their environment to an enormous extent. This extraordinary environmental impact has also made elephants a significant "indicator" species of the health and vitality of the African grassland. Elephants range widely over great distances and habitats, moving almost continuously, sleeping only four or five hours a day, and consuming many different types of vegetation. Thus elephants constitute an "umbrella" species, as well, for their conservation results in the protection of many other species. Finally, elephants may be the archetypal "flagship" species, for their aesthetic, symbolic, and utilitarian value has made this species the basis of legend, myth, and story in many civilizations throughout history.

A second facet of the elephant's unusual and often intense relationship with people has been its extraordinary similarity to humans. Elephants, like people, are extremely intelligent. They possess a wide range of vocalizations, a considerable learning capacity, an excellent memory, and an ability to engage in rational decision making, even passing on cultural norms from one generation to another. Similar to humans, elephants possess a complex social life, living in relatively large groups, maintaining strong family bonds, exercising considerable care and altruism, and being both organized and individualistic. Elephants and humans are alike in being long-lived creatures, too, with a slow reproductive biology, an extended childhood and adolescence, late puberty, and breeding over much of their life. Finally, humans and elephants are among the few creatures who can fundamentally alter and transform their environment. These various analogues between people and elephants help to explain why elephants have achieved such a special emotional and intellectual niche in the human imagination. More ominously, these common characteristics help to elucidate the extreme difficulty humans and elephants have had in coexisting in the same area, with the possible exception of those rare circumstances when people practice a largely hunter-gatherer or nomadic pastoralist way of life.

The recent decline of elephants reflects an accelerating process of competitive exclusion caused by the vast growth in human numbers, range, and technology. The African elephant's demise has been mainly a consequence of the reduction and fragmentation of its habitat and an excess of killing. Ultimately the diminishment of elephant habitat will probably determine its fate, but the more immediate threat to this animal's survival is its exploitation in connection with the international wildlife trade. Excessive killing of African elephants has been facilitated by the spread of modern weaponry, political instability, and, above all, the market for elephant ivory.

Elephant ivory has been a valuable commodity for thousands of years, used in an astonishing variety of products, and stored like gold as a form of wealth because of its rarity and durability. Prior epochs of excessive elephant exploitation have occurred—most infamously during the era of African exploration and the associated slave trade—but recent elephant exploitation may be unrivaled in its degree of wasteful, cruel, and destructive killing, particularly during the decades from 1970 to 1990. Many factors fostered this situation, including various economic, cultural, and technological forces that together increased the demand for elephant ivory and the capacity to market it. The elephant ivory trade especially developed in the Far East, as newly affluent nations like Japan and Taiwan purchased huge quantities of ivory mostly to manufacture signature seals, a product once restricted to the upper classes.

The effects of modern transportation, the availability of advanced weaponry, considerable civil unrest, the collusion of corrupt officials—all resulted in unprecedented levels of legal and illegal killing of elephants. The carnage led to the death of some 70,000 to 90,000 elephants during the worst years of the 1970s and 1980s. The African elephant population plummeted in half from an estimated 1.3 million animals in 1980 to some 650,000 by 1990. The average weight of an elephant tusk declined from 21 to 13 pounds, indicative of major impacts on the age structure of the species.

International outrage quickly developed, particularly among the Western nations. Many believed the species could be saved only by entirely eliminating the ivory trade. The United States Congress, responding to a rising chorus of public protest, passed an Elephant Protection Act in 1988, prohibiting the importation of elephant products. A nearly total worldwide ban on the trade in elephant ivory followed in 1990, promulgated by the Convention on International Trade in Endangered Species of Flora and Fauna (CITES), an agreement involving more than 120 nations. The CITES ban had an immediate effect: the price and consumption of elephant ivory dropped dramatically. With fewer market incentives, increased trade controls, and more vigilant law enforcement, poaching greatly diminished and, once again, elephant populations expanded. Certain southern African nations, however, objected strenuously to restrictions on the elephant trade, particularly Zimbabwe, Botswana, and South Africa. These countries claimed their better management practices had resulted in abundant and sometimes excessive elephant populations. They also contended that elephants seriously competed with their citizens for land and resources, damaged agricultural crops, and inflicted numerous human injuries and even deaths each year. Abundant elephant populations, effective management controls, and a cultural tradition of consuming wildlife, they argued, constituted sufficient justification for allowing the limited and sustainable harvest of elephants.

The current elephant controversy provides a powerful case study of the current debate for and against commercial wildlife utilization. Opponents of elephant harvesting and other forms of wildlife trade claim that the huge profits involved inevitably result in unsustainable exploitation and species decline. They also contend that managing for single species and single uses frequently results in ecologically unwise practices, among them the elimination of wild competitors and manipulation of habitats to produce unnatural surpluses. These skeptics further believe that the overriding emphasis on economic returns typically encourages unethical and cruel harvesting strategies. Resuming the global trade in elephant ivory, they say, would once again precipitate uncontrollable poaching of elephants—particularly in those African countries with limited capacities for monitoring and restricting the trade. They emphasize that the ban on elephant trade has been remarkably effective, even though it has been in existence for only a short time, and it would be both foolish and premature to rescind it.

Opponents of wildlife trade have also expressed serious doubts about the presumed economic benefits of wildlife utilization. They note most indigenous peoples in Third World countries obtain very limited economic returns from exploiting wildlife, most of the profits accruing to a small number of entrepreneurs, government officials, and foreign operators. These economic elites frequently alienate local villagers from the wildlife resource, who eventually come to view the commercial species management and protection as a choice of animals over people, particularly when the objective is to produce goods for a luxury trade to benefit wealthy foreigners. Finally, opponents of the wildlife trade often view as ethically repugnant the killing of an animal like the elephant distinguished for its extraordinary intelligence, complex social life, and presumed capacity for pain and suffering.

In contrast, advocates of elephant harvesting and wildlife trade cite the many benefits of commercial utilization, suggesting their opponents are naive, ignorant, and ethnocentrically biased. They further argue that practical wildlife use remains a necessity for most Third World countries mired in poverty and dependent on extracting natural resources as their primary source of foreign exchange. The income that could be generated, for example, from the sustainable harvesting of 15,000 African elephants each year might be in excess of $100 million annually.

Advocates of elephant harvesting also contend that the gravest threat to biological diversity in Africa is not the commercial trade but the conversion of wildlife habitat into agriculture, forestry, mining, roads, and other forms of large-scale development. They believe the survival of large and wide-ranging species like the elephant depends on its ability to outcompete economically other land uses. They say elephant utilization can produce large enough profits to provide a significant economic incentive for maintaining natural habitats. Proponents further contend that elephant utilization represents a more compatible and respectful acknowledgment of traditional African values. By contrast, they say elephant preservation constitutes an alien and imposed Western cultural bias.

Trade advocates finally claim that many objections to elephant harvesting represent technical problems amenable to solution through better management, improved science, modern technology, and more cooperative institutions involving local peoples and government officials.

This issue clearly represents a complex problem not easily resolved. Any form of commercial wildlife utilization must be compatible with three major principles involving wild living resources. First, all wildlife values of every major stakeholder should be equitably considered, avoiding any undue stress on a narrow set of benefits and interest groups at the expense of others. Commercial wildlife exploitation by definition emphasizes utilitarian and dominionistic values, but this focus should not preclude the recognition of equally important ecologistic, naturalistic, aesthetic, moralistic, humanistic, and symbolic benefits derived from an abundant and healthy resource. Sustainable wildlife management means a continuous stream of economic as well as other physical, emotional, intellectual, and spiritual values for the years ahead.

Second, wild living resource utilization should acknowledge the prevalence of widespread ignorance and uncertainty about species and ecosystems. This absence of full knowledge suggests that commercial wildlife exploitation should proceed with considerable caution and humility, recognizing the probability of the unexpected and unknown occurring. Notions of maximum sustainable yield should be abandoned in favor of the more elusive but more desirable goal of optimum sustainable yield. This harvesting objective means establishing population buffers allowing for our ignorance and uncertainty, as well as considering a species' contribution to its biological community and associated ecosystems. This management approach also emphasizes species and habitat interdependence across landscapes in place of a short-term focus on maximum production of single species.

Third, and finally, sustainable wildlife utilization should involve a sense of gratitude, respect, and reverence for the bounty nature provides people and society. Wildlife use represents a privilege, not a right, with attendant obligations placed on those who materially benefit from the exchange. Profiting from commercial wildlife should oblige people to engage in both biologically and ethically exemplary behavior.

Administration

These principles of wildlife utilization lead to the chapter's final focus: the administration of wildlife management. Earlier, in discussing endangered species protection, land use, and animal utilization, we noted significant shortcomings in traditional approaches to wildlife management and administration. Sometimes these problems reflect the unavoidable consequences of limited knowledge and unpredictability. Beyond these difficulties, however, various facets of contemporary wildlife management itself appear to be problematic.[12]

In discussing endangered species conservation, I suggested that many problems of wildlife management stem from questionable bureaucratic and organizational structures. Most public agencies continue to be unduly influenced by strong historical allegiances and traditional ideologies and limited by an inability or unwillingness to consider all relevant wildlife values in formulating and implementing policy. This deficiency has encouraged excessive predator and animal damage control, overexploitation of certain species, an undue emphasis on game animals, and a blemished record of biodiversity and endangered species conservation. These and other problems of management are, of course, much easier to identify than to rectify. Discussions of this kind can overlook the many sincere and significant efforts of scientists and policymakers, as well as the various technical, economic, and political constraints that impede needed change. Widespread problems and shortcomings, nonetheless, compromise the performance of many wildlife agencies. Yet many of these difficulties can be ameliorated through organizational and administrative reform.

A major obstacle to improvement continues to be limited understanding of complex organizations, particularly the behavior of government bureaucracies, and how to make these institutions perform more effectively. Organizational behavior constitutes alien terrain for most resource managers. By contrast, private corporations long ago recognized that their competitiveness depends as much on their organizational structure as on product engineering and marketing. Unlike private companies, most government agencies lack the incentive of a competitive environment or the bottom line of profits and losses for measuring performance. If biodiversity conservation is to succeed, wildlife agencies must develop a commitment to understand and improve their organizational structures, institute clear performance criteria, and create a more competitive environment.

Protecting wildlife will involve extensive consideration of the many internal and external factors that shape organizational behavior. This means confronting such institutional problems as inappropriate reward structures, conflicting administrative goals, limited competencies, inconsistent agendas, rigid leadership patterns, fragmented decision making, poor communication procedures, and inadequate accountability. Wildlife agencies can no longer afford to rationalize their limitations by blaming inadequate scientific knowledge, public irrationality, political pressure, and economic and social forces beyond their control when the trouble lies in the structure and dynamics of their own organizational bureaucracies.

Correcting problems of agency performance and behavior will be neither simple nor easy. The four suggestions offered here represent just a start toward trying to understand and address the many institutional problems encountered in the wildlife and natural resources field today. First, wildlife agencies should rely less on rigid standard operating procedures and more on flexible problem-solving approaches. Most biodiversity conservation problems are extraordinarily complex and vary greatly depending on location and circumstance. To cope with this variability requires an ability to respond to change and a management emphasis that tailors programs to dynamic situations and contexts.

Second, wildlife agencies need to become more accountable and measure their effectiveness by explicit performance criteria such as the number of species saved, biodiversity protected, and habitat restored. Clearly defined objectives and regular review procedures should improve organizational learning, corrective feedback, and institutional renewal. It might be worth experimenting with the privatization of certain management functions, such as endangered species recovery programs, to increase the competitive incentives for more effective and efficient action.

Third, wildlife management agencies should rely less on hierarchical command and control procedures. Agencies might employ more team-oriented, participatory, and consensual techniques. Rigid leadership and authoritarian decision making often discourage innovation, responsiveness to change, and new conservation approaches. Hierarchical structures also tend to reinforce established norms and traditional patterns of behavior. The concentration of power at the highest levels can result in unduly politically oriented leadership, an inordinate focus on bureaucratic behavior at middle management levels, and an unfortunate concentration of technical skills at the lowest levels of decision making. This separation of functions often blocks effective communication and the best use of scientific and professional skills and knowledge.

Finally, wildlife agencies should make their programs more relevant to the needs and perceptions of local communities. A local emphasis would seek to develop increased incentives for ordinary citizens to conserve and protect biodiversity. Current wildlife management approaches tend to concentrate power in centralized bureaucracies. This tendency often fosters resentment among local communities and regions toward distant and indifferent government agencies. This concentration of control frequently results in insufficient motivation for local people to manage their resources sustainably. Wildlife agencies need to develop co-management procedures and institutions that seriously involve local people in policymaking, while employing professionals to supply technical skills and emphasize the broad public interest. Biodiversity management should rely less on focusing power in the state and more on encouraging ordinary citizens to take constructive action.

Contemporary wildlife management should also pursue the goals of managing for both biological and sociological diversity. Managing for biological diversity means administering for all species, game or nongame, commodity or noncommodity, aesthetically appealing or unattractive, culturally significant or historically irrelevant. These distinctions increasingly possess little ecological or ethical sense. They constitute negative contrasts allied less with contemporary reality than with obsolete concepts of a species' presumed utility. These biases might once have served a useful purpose in deciding to manage for one species over another. But expanded knowledge of natural processes, the complexity of human dependence on wildlife, and the enormity of the current biodiversity crisis have rendered these assumptions not only antiquated but potentially harmful. Species once considered useless are increasingly recognized as possessing considerable ecological and even commodity value as human knowledge expands and the enormous complexity and potential utility of the biotic enterprise is more fully appreciated. Obscure organisms are more and more acknowledged for their contributions to such life-sustaining processes as nutrient cycling, plant reproduction, decomposition, and a host of other food and energy relationships. The accelerating scale of the biodiversity crisis suggests the need to manage for all species, game or nongame, charismatic megavertebrates or the countless invertebrates who help motor the world.

Wildlife agencies should move from a traditional focus on harvestable surpluses of favored species to the conservation and restoration of entire biological communities and ecosystems. This change does not mean abandoning the management of single species possessing significant utilitarian or cultural importance or those in peril of extinction. It does suggest, however, avoiding a focus on single species at the expense of other creatures or seeking to suppress or eliminate entire categories of life because they conflict with preferred species.

Managing for biological diversity further means a broadening of spatial and temporal horizons. Maintaining species interconnections at varying levels of biological organization requires managing for populations and habitats in an ecosystem and landscape context. This broad biogeographical perspective suggests the need to preserve ecological structure and complexity (maintaining wetlands, preserving all vegetational age classes, and the like) and ultimately ensuring the adequate protection of energy and metabolic output in natural systems. Adequate linkages, corridors, and connections among biological communities will also be important in seeking to avoid habitat fragmentation and degradation of ecological processes and dependencies. Temporally managing for biological diversity means considering human impacts on wildlife, not one species or one event at a time, but from a cumulative long-term perspective. This long-term view requires shifting from single-year objectives to strategic planning over many years, even attempting to maintain evolutionary processes across geological time scales.

These spatial and temporal goals, of course, represent exceedingly difficult and idealistic management objectives. Humans, perhaps like any creature, are parochial by nature; our attitudes and activities are deeply rooted in the interests of the moment. Managing for biological diversity may require new concepts of human well-being stretching beyond the limits of immediate experience. We may need to cultivate a view of the human species as both the consequence and the creators of evolution. This expanded concept might well engender a more compassionate humanity—a humanity more cognizant of its connections and impacts on the vast diversity of life.

Managing for sociological diversity represents another formidable challenge. As noted, it is essential to manage for all wildlife values. Most wildlife agencies, however, fail to incorporate values of importance to nontraditional users and interest groups. This omission has fostered controversy and conflict, particularly when deeply held values have been denied or even ridiculed. If wildlife agencies ally themselves with a narrow range of interests, they will inevitably be viewed as the captive of these constituencies and isolated from major segments of society. Managing for sociological diversity does not mean trying to please every group or clientele or forsaking the necessity of sound scientific judgment. It does suggest, however, understanding and incorporating wherever possible all critical wildlife values held by key groups in society. Public wildlife managers should be viewed as impartial agents whose primary objectives are to maintain the health of the resource, marshal scientific evidence, and serve as neutral mediators among competing constituencies.

Managing for sociological diversity will also demand an expanded recognition of the human dependence on the diverse experience of wildlife. This book has emphasized how people derive aesthetic, dominionistic, ecologistic, humanistic, moralistic, naturalistic, symbolic, and utilitarian benefits from living diversity and natural systems. The widest affiliation with nature and wildlife has been depicted as conferring essential physical, emotional, intellectual, and spiritual advantages. Managing for sociological diversity means providing people with opportunities for meaningfully and sustainably experiencing all these values as a basis for a rewarding life. These wildlife values vary according to demography, activity, and culture, and managing for sociological diversity means recognizing this heterogeneity. Varying wildlife values among demographic and activity groups should be considered in developing effective and equitable management programs. The entire public, rather than particular interest groups, constitutes the core clientele, underscoring the task of wildlife management as fundamentally one of people management.

Managing for both biological and sociological diversity will require a shift toward what might be called integrated wildlife management. This term suggests three broad goals. First, biophysical management objectives should be connected with social goals. This linkage views long-term wildlife sustainability as depending on the maintenance of an array of basic ecological and valuational processes. Biophysical criteria include preserving species diversity, overall organic matter or biomass, ecological structure and complexity (types and distributions of habitats and ecosystems, vertical and horizontal landscape heterogeneity), and the energy production of natural systems (nutrient flows, metabolic activity, biogeochemical cycles). Valuational sustainability means maintaining the capacity of natural systems to produce a wide range of basic benefits people derive from the diversity of life.

Second, integrated wildlife management means recognizing the need for interdisciplinary understanding—that is, acknowledging the indispensability of both the natural and social sciences in designing and implementing effective policies for conserving biodiversity. And third, integrated management means recognizing the complexity of the wildlife policy process. Managers must learn to handle a wide array of biophysical, socioeconomic, valuational, and organizational variables. They will also need to appreciate the often competitive character of most wildlife issues. Finally, managers will need to view wildlife policy as a dynamic process presenting different opportunities and constraints depending on the stage of its formulation and implementation.

The goal of integrated wildlife management will also necessitate fundamental shifts in the culture of the wildlife profession. Some of these changes have already been noted—the move from an emphasis on single species to all wildlife and associated ecosystems, for example, and a corresponding change from utilitarian and consumptive use values to all the wildlife perspectives held by varying groups in an increasingly pluralistic society. Most important, wildlife agencies must aspire to be the primary defenders of biological diversity rather than simply functioning as service providers to narrow and traditional clienteles. This recommendation is not meant to denigrate the historic importance of hunters and fishers, who have over the years provided much of the moral and financial support for America's largely successful system of wildlife management. Financial dependence on taxing sportsmen, however, has distorted the goals of the profession by encouraging the wildlife management field to accommodate itself to the needs of this clientele rather than the condition of wildlife more generally. The wildlife management profession must break the financial ties that bind and emerge from the narrow political cul-de-sac in which it has entrapped itself. The funding basis for wildlife management should be altered and broadened.

Conclusion

The policy and management recommendations offered here would take many years to implement through a process of incremental change. Although this gradualist approach might be less politically challenging and more likely to succeed, the scale of the current biodiversity crisis demands more fundamental and immediate institutional change. The organization and leadership of wildlife management appear to call for basic structural reform. Management agencies will need to marshal all their resources and expertise to arrest the current trends toward ecological decline and sociological alienation from nature and living diversity.

CHAPTER 9

Education and Ethics

THE TASK of conserving biological diversity may constitute one of the most critical challenges the human species has ever faced. The stakes may include our future material and physical well-being, as well as the probability of leading lives replete with emotional, intellectual, and spiritual value. The best of society's scientific knowledge, political skills, and organizational abilities will be required to accomplish this task.

Winning the hearts and minds of the general public and its leadership will be a necessity. People will need to be informed and convinced of how much humanity continues to depend on intimate, diverse, and satisfying affiliations with nature and living diversity. People will need to rekindle their capacity for experiencing wonder, inspiration, and joy from contact with the natural world and its many creatures.

Education and ethics, therefore, will be as important as science, policy, and management in attaining this goal.[1] A knowledgeable and ethically responsible citizenry—environmentally literate and morally concerned—will be an indispensable ingredient in securing and restoring the integrity and health of the biosphere. This final chapter explores some aspects of the need to educate people about the value of living diversity—and the importance of cultivating an ethic based on an expanded notion of personal and social self-interest.

Education

Emphasizing the virtues of environmental education represents a far easier task than its implementation. Rallying for environmental literacy has become something of a cliché of late, pronounced with great solemnity and at times conviction. The sad reality, however, remains: environmental education receives far less financial support or professional prestige than either natural resource science, policy, or management. Limited training and staff are typically allocated to the task of environmental education and, all too often, public relations and mass media efforts have become confused with the far more difficult challenge of transforming people's feelings, knowledge, and beliefs about biological diversity and conservation. Most environmental education programs have limited resources, few well-trained personnel, and little status. Methodologies for educating people about living diversity and its role in human life tend to be inadequate, inconsistent, and lacking in sufficient understanding of the normal learning process or how to tailor programs to varying age and demographic groups.

Many armies march under the banner of environmental education. Informing the public about nature and living diversity has ranged from serious undertakings intended to instill a deep and abiding appreciation of the natural world to little more than public relations efforts seeking to cultivate organizational and political allegiance. Trying to obtain support for an agency's programs or financial backing for certain groups is not an unworthy goal, but it can become problematic when efforts to manipulate people for largely self-serving purposes are substituted for more lasting and meaningful education. Public relations and mass media efforts can never replace the more complex and relevant task of nurturing a knowledge and conviction among people of the indispensable role of a healthy and diverse natural environment. Biodiversity education should seek to inform people, emotionally and intellectually, about the role of the living environment in their lives. Moreover, this knowledge should be delivered in a variety of formal, informal, and indirect educational contexts. Each of these objectives will be briefly examined.[2]

People need to learn about the connection between human life and the health and abundance of the natural world not just cognitively but emotionally and in value terms as well. Contemporary society lacks adequate appreciation of how much human existence depends on varying interactions with biological diversity to achieve lives of physical, emotional, and intellectual meaning. Many people assume the great triumphs of modern life such as the conquest of hunger, disease, and material want have resulted from suppressing and eliminating other life forms. Biodiversity education must dispel this deep disconnection from the natural world. In its place it must instill the feeling, knowledge, and belief of how much human sustenance and spiritual enrichment depends on maintaining a rich variety of relationships to nature and living diversity.

Affective learning means cultivating an emotional appreciation of how the living world offers people profound opportunities for kinship, wonder, and beauty. These sentiments can be unlocked through building attachments to the familiar and appealing in nature, evoking both empathy and loyalty. With sufficient effort, nearly any aspect of the living world can become a meaningful and satisfying source of emotional bonding, providing a nearly limitless opportunity for affective learning, pleasure, even enchantment.

Cognitive learning, by contrast, means acquiring factual and conceptual knowledge of living diversity and its importance to human well-being. This intellectual education often starts by instructing people in identifying and classifying various elements of the natural world. Eventually cognitive instruction moves to more conceptual understanding of environmental structure, function, and process. Intellectual education is complete when it develops an awareness of how humans impact the living world, as well as the skills necessary for exercising responsible stewardship and intervention.

Values education constitutes a synthesis of cognitive and affective learning. This education focuses on attitudes and beliefs consistent with a deep appreciation of the role of living diversity in human life. It begins by understanding the many ways natural systems render human existence more materially and physically secure. This emphasis eventually expands to include an appreciation of how living abundance and diversity also nourish the human capacity for exploration, imagination, and creativity. Values education concludes with a deep ethic of care and compassion for all life and an abiding commitment to ensure the natural world's healthy perpetuation.

The typology of values cited throughout the book represents a template for organizing the task of affective, cognitive, and evaluative education. Each value tends to emphasize a certain learning level. Although each of the values may be expressed at all three learning levels, aesthetic, humanistic, and negativistic values tend to be most often manifest emotionally. Moralistic, utilitarian, and dominionistic values, however, tend to be frequently expressed and therefore learned as broad belief orientations. Finally, ecologistic, naturalistic, and symbolic values are essentially cognitive.

Biodiversity education should take place in formal, informal, and indirect learning contexts. These diverse instructional settings offer the greatest opportunity to communicate with the largest number of people in the widest possible ways. Each educational setting presents its own challenges and opportunities, requiring different methodologies for teaching an understanding and appreciation of nature and biological diversity.

Formal education typically occurs in schools and organized classrooms. It tends to be most effective when focusing on cognitive learning, particularly the science of natural process, structure, and taxonomy. This often complex form of education frequently requires close supervision, evaluation, and repetition in a structured setting. The classroom usually provides the most suitable context for complex sequential learning of formally organized curricula iteratively moving from simple to more complex understandings of biophysical and environmental functioning.

Informal education involves less structure. It tends to be voluntary and self-paced and often dispenses with formal testing or other rigid controls over learning. Zoological parks, botanical gardens, natural history and science museums, national parks and other protected areas, nature centers, outdoor education programs—all provide the setting for informal environmental education. Informal learning frequently relies on eliciting people's interest and attention, if not affection, and indeed sometimes stresses entertainment as much as education. A frequent challenge of informal education is to ensure entertainment does not undermine learning. Informal education has an extraordinary potential for stimulating affective and, to a lesser degree, cognitive understanding of nature and living diversity. Zoos, nature centers, national parks, and other informal settings frequently provide unrivaled opportunities for nurturing a deep appreciation for the natural world. Moreover, the intimate experience of nature in outdoor settings can sometimes inspire profound feelings of connection with nature and living diversity.

Indirect environmental education focuses on the role of the mass media, particularly newspapers, magazines, radio, television, and film. The experience is indirect because no contact typically exists between the source of the information and the recipient. An unfortunate consequence of this lack of direct relationship is the frequent assumption that the providers of the information have no particular educational obligation or responsibility.[3] Few environmental journalists and filmmakers view themselves as needing special training or knowledge about nature and living diversity. And, few regard themselves as particularly accountable for the accuracy, depth, and quality of the information conveyed. Educating the public about the natural world, however, will necessitate a far more serious commitment from the mass media. The discussion of wildlife film and television in an earlier chapter clearly suggests the powerful impact of indirect learning on people's knowledge and attitudes about nature and living diversity. An environmentally informed public will require a responsible and innovative media striving to accomplish the difficult fusion of mass communication with education.

Ethics

Although an informed, aware, and appreciative public is indispensable, it is not enough. Just as important is the need for cultivating an ethic of care and compassion for the diversity of life.[4] Historic disregard and disrespect for other species was considered earlier in the book, particularly in discussing past persecution of wolves, whales, and other creatures. This callous destruction of nonhuman life has become more a facet of this country's past than its present. Even so, modern society continues in more silent and less obvious ways to destroy a substantial fraction of life on earth—perhaps to a greater degree than at any time in human history. Moreover, the average person remains but dimly aware of the scale of this impact, experiencing little burden, guilt, or sense of responsibility for correcting this devastating process. We can no longer afford to accept this degree of ignorance and indifference. The fabric of planetary life is under siege as vast expressions of creation are ripped from their evolutionary moorings by varying combinations of greed, arrogance, and apathy. Thousands of singularly distinctive species, each a unique expression of millions of years of adaptational travail, oblige us to devote whatever wisdom and ethics we can to the task of slowing and then reversing this tide of ultimately self-defeating destruction. We need to alter what, in our collective insanity, we have come to regard as normal.

An ethic of care and concern for living diversity can be attained through an expanded concept of personal and social self-interest.[5] The impoverishment of life on earth reduces the human potential for achieving meaningful and satisfying lives physically, emotionally, and intellectually. Degrading nature constitutes more than material harm: it encompasses as well a profound loss of psychological bearings and the debasement of the human experience. Our chance to lead fulfilling lives springs from an array of affiliations and dependencies on living diversity. The natural world remains at the core of our species' physical and mental being, and human security depends on maintaining a variety of intricate and subtle relations with the vast matrix of life. People are rooted in natural, not artificial, realities: human personality and society continue to rely on a myriad of relations with the earth and its creatures. The human capacities for caring, coping, curiosity, communication, harmony, kinship, order, and more remain contingent on intimate affiliation with the diversity of life.

These physical and psychological benefits constitute the self-interested basis for an ethic of care and concern for biological diversity. Through celebrating and protecting life, we may achieve a renewed sense of personal and social worth. This broad basis for rationalizing a deep ethical commitment to maintaining the health and abundance of living diversity is suggested by Edward Wilson:

> What humanity is doing now in a single lifetime will impoverish our descendants for all time to come. Yet critics often respond, "So what?" . . . The answer most frequently urged . . . is . . . material wealth . . . at risk. . . . This argument is demonstrably true . . . but it contains a dangerous practical flaw. . . . If . . . judged by . . . potential material value, [species] can be priced, traded off against other sources of wealth, and—when the price is right—discarded. . . . The species-right argument alone, like the materialist argument alone, is a dangerous play of the cards. . . . The independent-rights argument, for all its directness and power, remains intuitive, aprioristic, and lacking in objective evidence. . . . A simplistic adjuration for the right of a species to live can be answered by a simplistic call for the right of people to live. . . . A robust and richly textured anthropocentric ethic [can instead be made] based on the hereditary needs of our species . . . [for] the diversity of life has immense aesthetic and spiritual value.[6]

As Wilson suggests, an ethic based on materialistic or altruistic arguments alone fails to capture the full range of values and benefits humans derive from the diversity of life. Preventing an animal's pain, harm, and suffering is commendable, but this ethical perspective suffers from its narrow applicability. For most people, preventing harm to other creatures depends on their ability to empathize with species much like ourselves, mainly the higher vertebrates. The great mass of life and the ecosystems that make them possible are commonly omitted from this moral consideration.

Some advocate a more encompassing ethic viewing all creatures as possessing an intrinsic right to exist. Species extinction from this perspective constitutes the worst and most wanton form of killing: destroying birth itself by eliminating the possibility of a distinctive life form ever replicating itself. This ethical argument remains flawed, however, by its limited power to convince more than a very few people of the virtue of denying their own self-interest for the sake of a highly abstract notion of awarding all species an inalienable right to exist.

Wilson notes another rationalization for an ethic of species protection: the actual and potential material advantages derived from other organisms. This utilitarian argument stresses the practical benefits people obtain from living diversity now and in the future as human knowledge expands to exploit the vast genetic wealth of the natural world. This argument has serious limitations, however. The most significant is the relatively few number of species that will ever reasonably improve human material well-being. The seeds of self-destruction are sown in any species protection position that extends value to only a portion of life—declaring, by implication, the remainder expendable. It seems dubious, too, to place much confidence in an ethic whose materialistic logic produced the biodiversity crisis in the first place. Even so, what makes the utilitarian argument so persuasive is its appeal to personal and social self-interest. Rather than relying on people's limited ability to empathize with nonhuman life or viewing all species as possessing intrinsic rights independent of human welfare, the utilitarian perspective argues for protecting living diversity because it offers the promise of a richer and more rewarding life.

The position advocated here expands the utilitarian concept beyond the simply material. The many values of living diversity reflect the varied ways natural systems render possible not just human physical but also emotional, intellectual, and spiritual well-being. Nature continues to shape our values, define our culture, and offer the possibility for experiencing beauty, meaning, and purpose. The values described in this book reflect the many ways the living world can enhance material wealth, empirical truth, aesthetic worth, intellectual inspiration, emotional bonding, physical skill, spiritual solace, and much more. A broad anthropocentric ethic of life draws nourishment from recognizing how much humans depend on a world of natural abundance and variety.

Biological richness may seem contrary to conventional notions of wealth. Abundance, however, occurs at least economically, culturally, and naturally. Economic wealth tends to be measured monetarily, cultural wealth by a society's arts and customs, often considered beyond price, while natural wealth reflects the richness of the nonhuman world. These may constitute separate indices of wealth, but each ultimately depends on its relationship to the others. No society can retain for long its economic or cultural prosperity if it is built upon a despoiled natural world.

The ethical challenge is to expand our understanding of how human existence derives sustenance and spirit from the richness of its connections with the vast diversity of life. We must dispel the great fallacy of the modern age that human society no longer requires varied and satisfying connections with the nonhuman world. One reason for the massive degradation of life on earth is that people have somehow come to believe biological diversity constitutes so much excess to squander.

Perhaps it is no coincidence that the current maelstrom of biological destruction has occurred during possibly the most violent century in human history. At the root of the modern discontent may be the denial of our deep craving for meaningful association with the rest of creation. Perhaps we consign ourselves to a deep and threatening aloneness when we choose to stand apart and destroy the very processes that offer the potential for emotionally, intellectually, and spiritually meaningful lives. As Richard Nelson suggests, our "world view isolates us from the natural community and leaves us spiritually alienated from nonhuman life. . . . We have created a profound and imperiling loneliness."[7]

Recognizing our profound dependence on living diversity may carry the promise of more aesthetically attractive, ecologically productive, and psychologically satisfying lives. This does not mean returning to some bucolic, self-sufficient immersion in nature. Modern technology and human population numbers preclude this possibility. Much of society's energy, food, and material needs will no doubt continue to be derived from the large-scale production of goods transported over long distances. Humans will remain competitive with other life forms as an inevitable consequence of the struggle to survive and prosper. Utilizing nature's abundance, however, can proceed with respect and understanding of the many physical, emotional, and intellectual ways we depend on a sustainable, healthy, and diverse biota.

The willingness to coexist with the rest of creation should enhance rather than diminish the human condition. Our standing at the pinnacle of the great chain of being may be enlarged rather than lessened by better appreciating our varied connection with the diversity of life. As Wilson suggests: "The more we know of other forms of life, the more we enjoy and respect ourselves. Humanity is exalted not because we are so far above other living creatures, but because knowing them well elevates the very concept of life."[8]

The experience of life's varied enchantments can proceed anywhere humanity occurs: from the agricultural field to the small town to the urban landscape. People need only bring to these varied experiences a sense of wonder, appreciation, and joy for the living fabric. The more we plumb the depths of nature, the more we encounter its unrivaled capacity to nourish the human body and spirit. Every person possesses the ability to mine this creation and thereby enrich his or her existence. This represents the ultimate self-interest of an ethic of respect and reverence for the value of life.

NOTES

PROLOGUE

1. Scott McVay explores the many ways creatures can offer people profound opportunities for emotional and intellectual connection with the natural world in his "Prologue" to *The Biophilia Hypothesis,* edited by myself and E. O. Wilson (Washington, D.C.: Island Press, 1993). Edward O. Wilson offers similar insights in his book *Biophilia: The Human Bond with Other Species* (Cambridge: Harvard University Press, 1984). The notion of nature and animals as a "magic well" is derived from Karl von Frisch, as discussed in both McVay and Wilson.
2. This Ojibway quote is taken from publicity for the Geraldine R. Dodge Foundation's 1992 Poetry Festival held in Waterloo, New Jersey.

CHAPTER 1: INTRODUCTION

1. From S. Flader's insightful biography of Aldo Leopold, *Thinking Like a Mountain: Aldo Leopold and the Evolution of an Ecological Attitude Toward Deer, Wolves, and Forests* (Columbia: University of Missouri Press, 1974), p. 188.

2. Very little scientific study had occurred prior to the 1970s on people/animal relationships. For illustrations of early research see: J. Hendee, "A Multiple Satisfactions Approach to Game Management," *Wildlife Society Bulletin* 2 (1974):104–113; B. Levinson, *Pets and Human Development* (Springfield: Charles C. Thomas, 1972); K. Lorenz, *King Solomon's Ring* (New York: New American Library, 1952); R. Nash, *Wilderness and the American Mind* (New Haven: Yale University Press, 1976).
3. Some historical background of the U.S. Fish and Wildlife Service can be found in J. Tober, *Who Owns the Wildlife: The Political Economy of Conservation in 19th Century America* (Westport: Greenwood Press, 1981); T. Lund, *Wildlife Law* (Berkeley: University of California Press, 1980); M. Bean, *The Evolution of National Wildlife Law* (New York: Praeger, 1983); T. Clarke and D. McCool, *Staking Out the Terrain: Power Differentials Among Natural Resource Agencies* (Albany: State University of New York Press, 1985).
4. Persons at the Fish and Wildlife Service who were particularly helpful in launching the national study included Lynn Greenwalt, director of the service at the time and now at the National Wildlife Federation, and Dr. Lynn Llewellyn, who is still with the service.
5. Early versions of the typology of basic attitudes toward animals appeared in two of my publications: "From Kinship to Mastery: A Study of American Attitudes Toward Animals" National Association for the Advancement of Humane Education, 2(1976):24–37; "Contemporary Values of Wildlife in American Society," in W. W. Shaw and I. Zube, eds., *Wildlife Values* (Ft. Collins: U.S. Forest Service Rocky Mountains Forest and Range Experiment Station, 1980).
6. Three publications of E. O. Wilson have been particularly influential in my work: *Sociobiology: The New Synthesis* (Cambridge: Harvard University Press, 1975); *Biophilia* (Cambridge: Harvard University Press, 1984); and *The Diversity of Life* (Cambridge: Harvard University Press, 1992).
7. S. R. Kellert and E. O. Wilson, eds., *The Biophilia Hypothesis* (Washington, D.C.: Island Press, 1993).
8. This figure is discussed in detail in Chapter 2. For useful references, see note 51 for Chapter 2. This point is derived mainly from Wilson's *The Diversity of Life.*
9. H. Beston, *The Outermost House* (New York: Ballantine Books, 1971), p. 174. Beston also offered this equally compelling insight on page 19 of *The Outermost House:* "We need another and a wiser and perhaps a more mystical concept of animals. Remote from universal nature, and living by complicated artifice, man's civilization surveys the creature through the glass of his

knowledge and sees thereby a feather magnified and the whole image in distortion. . . . In a world older and more complete than ours they move finished and complete, gifted with extensions of the senses we have lost or never attained, living by voices we shall never hear. They are not brethren, they are not underlings, they are other nations, caught with ourselves in the net of life and time, fellow prisoners of the splendor and travail of the earth."

CHAPTER 2: VALUES

1. For the material and economic value of biological diversity see N. Myers, *The Sinking Ark: A New Look at the Problem of Disappearing Species* (Oxford: Pergamon, 1979) and *A Wealth of Wild Species: Storehouse for Human Welfare* (Boulder: Westview, 1983); M. L. Oldfield, *The Value of Conserving Genetic Resources* (Sunderland: Sinauer, 1989); C. and R. Prescott-Allen, *The First Resource* (New Haven: Yale University Press, 1986); B. Groombridge, ed., *Global Biodiversity: Status of the Earth's Living Resources* (London: Chapman & Hall, 1992); various essays in E. O. Wilson and F. M. Peter, eds., *Biodiversity* (Washington, D.C.: National Academy Press, 1988); J. G. Robinson and K. H. Redford, eds., *Neotropical Wildlife Use and Conservation* (Chicago: University of Chicago Press, 1991).
2. The importance of personally deriving practical benefits from nature is powerfully articulated by Aldo Leopold in his classic *Sand County Almanac* (New York: Oxford University Press, 1966) and in H. Rolston III, *Philosophy Gone Wild: Essays in Environmental Ethics* (Buffalo: Prometheus Books, 1986).
3. The sense of curiosity and discovery is described by Wilson in *Biophilia: The Human Bond with Other Species.*
4. These benefits are described by R. Ulrich in various essays including "Aesthetic and Affective Response to Natural Environment," in I. Altman and J. Wohlwill, eds., *Behavior and the Natural Environment* (New York: Plenum Press, 1983); "Stress Recovery During Exposure to the Natural and Urban Environments," *Journal of Environmental Psychology* 11(1991):201–230; "Biophilia, Biophobia, and Natural Landscapes," in *The Biophilia Hypothesis.* Another excellent reference is S. and R. Kaplan, *The Experience of Nature* (New York: Cambridge University Press, 1989).
5. G. Seilstad, *At the Heart of the Web* (Orlando: Harcourt Brace Jovanovich, 1989), p. 285; Ulrich, Aesthetic and Affective Response, p. 203.
6. Leopold, *Sand County Almanac,* p. 176
7. Ibid., p. 266.

8. Interactions and relationships between people and invertebrates are thoroughly discussed in Chapter 5 and in my paper, "Values and Perceptions of Invertebrates," *Conservation Biology* 7(1993):845–855.
9. See, for example, J. C. Ryan, "Life Support: Conserving Biological Diversity," *Worldwatch Report* 108(1992); M. Gadgil, "India's Deforestation: Patterns and Processes," *Society and National Resources* 3:131–143.
10. Ulrich, "Aesthetic and Affective Response to Natural Environment," 1983, p. 109.
11. Yi-Fu Tuan, *Passing Strange and Wonderful: Aesthetics, Nature, and Culture* (Washington, D.C.: Island Press, 1993); Y. Saito, *The Aesthetic Appreciation of Nature: Western and Japanese Perspectives and Their Ethical Implications* (Ann Arbor: University Microfilms, 1983).
12. A more thorough discussion of this point can be found in my essay, "The Biological Basis for Human Values of Nature," in *The Biophilia Hypothesis.*
13. Leopold, *Sand County Almanac,* pp. 129–130.
14. G. Schaller, *Stones of Silence* (New York: Viking Press, 1982).
15. Leopold, "Thinking Like a Mountain," in *Sand County Almanac.*
16. H. Rolston, "Beauty and the Beast: Aesthetic Experience of Wildlife," in D. Decker and G. Goff, eds., *Valuing Wildlife: Economic and Social Perspectives* (Boulder: Westview, 1987).
17. For two excellent essays on this topic with useful references see J. H. Heerwagen and G. H. Orians, "Humans, Habitats, and Aesthetics," and R. Ulrich, "Biophobia, Biophilia, and Natural Landscapes," both in *The Biophilia Hypothesis.*
18. Leopold, *Sand County Almanac,* p. 137.
19. See, for example, P. Shepard's seminal book, *Thinking Animals: Animals and the Development of Human Intelligence* (New York: Viking Press, 1978).
20. This point is associated with a number of anthropologists including C. Lévi-Strauss, *The Raw and the Cooked* (New York: Harper & Row, 1970) and *The Savage Mind* (Chicago: University of Chicago Press, 1966); E. Leach, "Anthropological Aspects of Language," in E. H. Lenneberg, ed., *New Directions in the Study of Language* (Cambridge: MIT Press, 1975).
21. E. Lawrence, "The Sacred Bee, the Filthy Pig, and the Bat Out of Hell: Animal Symbolism as Cognitive Biophilia," in *The Biophilia Hypothesis,* p. 300.
22. Shepard, *Thinking Animals,* pp. 249 and 2.
23. C. Jung, *The Archetype and the Collective Unconscious* (New York: Pantheon Books, 1959); B. Bettelheim, *The Uses of Enchantment* (New York: Vintage Books, 1977); J. Campbell, *Myths to Live By* (New York: Viking Press, 1973); and the previously cited works by Lévi-Strauss.

24. Shepard, *Thinking Animals*, p. 247.
25. Lawrence, "The Sacred Bee," pp. 336–337.
26. Some insights regarding this tendency are provided by J. Ortega y Gasset, *Meditations on Hunting* (New York: Macmillan, 1986); H. Rolston, "Values in Nature," in *Philosophy Gone Wild;* and S. and R. Kaplan, *The Experience of Nature.*
27. Rolston, *Philosophy Gone Wild,* p. 88.
28. This tendency has been noted in research focusing on programs involving personal development through the outdoors (such as Outward Bound and the National Outdoor Leadership School). Relevant discussions and references can be found in S. and R. Kaplan, *The Experience of Nature;* T. Hartig et al., "Restorative Effects of Natural Environment Experiences," *Environment and Behavior* 23(1993):3–26; S. Kaplan and J. F. Talbot, "Psychological Benefits of a Wilderness Experience," in I. Altman and J. Wohlwill, eds., *Human Behavior and Environment* (New York: Plenum, 1983); J. Hendee, G. H. Stankey, and R. C. Lucas, *Wilderness Management* (Washington, D.C.: U.S. Forest Service, 1978).
29. An extensive and informative literature has developed around the human/animal companion bond. Many articles in the journal *Anthrozöos* and publications of the Delta Society have discussed this bond. Specific references include A. Katcher and A. Beck, eds., *New Perspectives on Our Lives with Companion Animals* (Philadelphia: University of Pennsylvania Press, 1983); J. Serpell, *In the Company of Animals* (Oxford: Basil Blackwell, 1986); R. B. Anderson et al., *The Pet Connection* (Minneapolis: University of Minnesota Press, 1984); B. Fogle, *Interrelations Between People and Pets* (Springfield: Charles C. Thomas, 1981).
30. K. Thomas, *Man and the Natural World* (New York: Pantheon Books, 1983).
31. In addition to publications of the Delta Society and *Anthrozöos* noted above, this issue is considered in the essay by A. Katcher and G. Wilkins, "Dialogue with Animals: Its Nature and Culture," in *The Biophilia Hypothesis.*
32. Serpell, *In the Company of Animals,* pp. 114–115.
33. The reader could look at E. O. Wilson, *The Diversity of Life,* and B. Groombridge, ed., *Global Biodiversity,* for useful information on taxonomic distributions by varying species groups.
34. This point is made in E. O. Wilson's essay, "Biophilia and the Conservation Ethic," in *The Biophilia Hypothesis.*
35. See, for example, J. D. Hughes, *American Indian Ecology* (El Paso: Texas Western Press, 1983); B. Callicott, "Traditional European and American

Indian Attitudes Toward Nature: An Overview," *Environmental Ethics* 4(1989):293–318; A. H. Booth et al., "Ties That Bind: Native American Beliefs as a Foundation for Environmental Consciousness," *Environmental Ethics* 12(1990):27–43. See also the references cited in note 18 of Chapter 6.

36. See the fine essay by R. Nelson, "Searching for the Lost Arrow: Physical and Spiritual Ecology in the Hunter's World," in *The Biophilia Hypothesis,* as well as other writings by Nelson including *Hunters of the Northern Ice* (Chicago: University of Chicago Press, 1969) and *Make Prayers to the Raven* (Chicago: University of Chicago Press, 1983).
37. John Steinbeck, *Log from the Sea of Cortez* (Mamaroneck: P. P. Appel, 1941), p. 93.
38. Walt Whitman, *Leaves of Grass* (New York: Penguin, 1976); cited in N. Foerster and R. Falk, eds., *American Poetry and Prose* (Boston: Houghton Mifflin, 1960), p. 55.
39. See, for example, some of Aldo Leopold's remarks in *Sand County Almanac* regarding the link between conservation, ecology, and ethical integrity.
40. Wilson, "Biophilia and the Conservation Ethic," p. 39.
41. See, for example, A. Öhmans, "Face the Beast and Fear the Face: Animal and Social Fears as Prototypes for Evolutionary Analyses of Emotion," *Psychophysiology* 23(1986):123–145; Ulrich, "Biophobia, Biophilia, and Natural Landscapes," in *The Biophilia Hypothesis.*
42. Ulrich, "Stress Recovery," p. 206.
43. T. Schneirla, *Principles of Animal Psychology* (Englewood Cliffs: Prentice-Hall, 1965); T. Hardy, "Entomophobia: The Case for Miss Muffett," *Entomology Society America Bulletin* 34(1988):64–69; J. Hillman, *Going Bugs* (Gracie Station: Spring Audio, 1991).
44. B. Lopez, *Of Wolves and Men* (New York: Scribner's, 1978).
45. Wilson, *Biophilia.*
46. The processes of biocultural evolution are elucidated in C. J. Lumsden and E. O. Wilson, *Genes, Mind, and Culture* (Cambridge: Harvard University Press, 1981) and "The Relation Between Biological and Cultural Evolution," *Journal of Social and Biological Structure* 8(1983):343–359.
47. Wilson, "Biophilia and the Conservation Ethic," pp. 31–32.
48. R. Dubos, *Ecology and Religion in History* (New York: Oxford University Press, 1969), p. 129.
49. H. Iltis, keynote address, transactions of a symposium, *The Urban Setting: Man's Need for Open Space* (New London: Connecticut College, 1980), pp. 3 and 5.

50. Leopold, *Sand County Almanac*, p. 266.
51. Useful statistics on rates of species extinction and endangerment can be found in the following sources: E. O. Wilson, *The Diversity of Life;* B. Groombridge, ed., *Global Biodiversity;* G. O. Barney, ed., *Global 2000 Report to the President: Entering the 21st Century* (Washington, D.C.: Council on Environmental Quality, 1980); V. Ziswiller, *Extinct and Vanishing Animals* (London: English University Press, 1967); C. A.W. Guggisberg, *Man and Wildlife* (New York: Arco, 1970); E. Norse, *Global Marine Biological Diversity* (Washington, D.C.: Island Press, 1993); WRI/IUCN/UNEP, *Global Biodiversity Strategy* (Washington, D.C.: World Resources Institute, 1992); S. Wells, R. Pyle, and N. M. Collins, eds., *The IUCN Invertebrate Red Data Book* (Gland, Switzerland: International Union for the Conservation of Nature and Natural Resources, 1983); N. Myers, "Questions of Mass Extinction," *Biodiversity and Conservation* 2(1994):2–17; N. Myers, "Global Biodiversity II: Losses," in G. K. Meffe and C. R. Carroll, eds., *Principles of Conservation Biology* (Sunderland: Sinauer, 1994); L. Kaufman and K. Mallory, eds., *The Last Extinction* (Cambridge: MIT Press, 1993); W. L. Minckley and J. E. Deacon, eds., *Battle Against Extinction: Native Fish Management in the American West* (Tucson: University of Arizona Press, 1991).
52. Various publications describe the tragedy of Hawaii's endemic birds. A particularly informative book is A. J. Berger, *Hawaiian Birdlife* (Honolulu: University of Hawaii Press, 1981). See also the references cited in note 9 of Chapter 7.
53. See, for example, F. H. Bormann and G. Likens, *Patterns and Process in a Forested Ecosystem* (New York: Springer-Verlag, 1979); R. S. DeGroot, *Functions of Nature* (Groningen: Wolters-Noorhoff, 1992); C. J. Krebs, *Ecology* (New York: Harper & Row, 1978); R. M. May, *Stability and Complexity in Model Ecosystems* (Princeton: Princeton University Press, 1973).
54. P. Matthiessen, *Wildlife in America* (New York: Viking Press, 1989), p. 21.
55. Leopold, *Sand County Almanac*, p. 109.

CHAPTER 3: AMERICAN SOCIETY

1. The issue of resource and ecosystem valuation was recently the focus of a major U.S. Environmental Protection Agency/Conservation Foundation assessment. For valuable references on varying biological, economic, and social-psychological strategies for valuing nature see R. S. DeGroot, *Functions of Nature;* G. Peterson, B. L. Driver, and R. Gregory, eds., *Amenity Resource Valuation* (State College, Pa.: Venture, 1988); M. B. Usher, *Wildlife Conservation Valuation* (London: Chapman & Hall, 1986); D. J. Decker and

G. R. Goff, eds., *Valuing Wildlife: Economic and Social Perspectives* (Boulder: Westview, 1987); W. Westman, *Ecology, Impact Assessment, and Environmental Planning* (New York: Wiley, 1985); R. Constanza, B. G. Norton, and B. D. Haskell, eds., *Ecosystem Health: New Goals for Environmental Management* (Washington, D.C.: Island Press, 1992); M. Freeman, *The Measurement of Environmental and Resource Values: Theory and Methods* (Washington, D.C.: Resources for the Future, 1993); S. Kellert, "Assessing Environmental and Wildlife Values in Cost-Benefit Analysis," *Journal of Environmental Management* 18(1984):353–363.

2. A series of reports on the national study written by myself and various colleagues were issued by the Government Printing Office. Although these reports are no longer in print, they may be obtained by contacting the Office of Program Plans, U.S. Fish and Wildlife Service, Department of Interior. The reports include: *Public Attitudes Toward Critical Wildlife and Natural Habitat Issues; Activities of the American Public Relating to Animals; Knowledge, Affection, and Basic Attitudes Toward Animals in American Society* (with J. Berry); *20th Century Trends in American Perceptions and Uses of Animals* (with M. Westervelt); *Children's Attitudes, Knowledge, and Behaviors Toward Animals* (with M. Westervelt).
3. S. Kellert and M. Westervelt, *20th Century Trends in American Perceptions and Uses of Animals.*
4. Historical trends, a complex subject, are discussed in relation to particular species in other chapters. For a useful general historical understanding of wildlife and nature in America see P. Matthiessen, *Wildlife in America* (New York: Viking, 1989); T. Dunlap, *Saving America's Wildlife* (Princeton: Princeton University Press, 1989); J. B. Trefethen, *The American Landscape: 1776–1976* (Washington, D.C.: Wildlife Management Institute, 1975); D. Worster, *The Wealth of Nature: Environmental History and the Ecological Imagination* (New York: Oxford University Press, 1993); J. M. Petulla, *American Environmental History: The Exploitation and Conservation of Natural Resources* (San Francisco: Boyd & Fraser, 1977); M. Oelschlaeger, *The Idea of Wilderness* (New Haven, Yale University Press, 1991).
5. For useful readings regarding the work of Jean Piaget, Lawrence Kohlberg, and other learning and developmental theorists and scientists see C. June Maker, *Teaching Models of the Gifted* (Austin: Pro-ed, 1982); L. Kohlberg, *Essays on Moral Development* (San Francisco: Harper & Row, 1984); J. Piaget, *Judgment and Reasoning in the Child* (New York: Littlefield, Adams, 1928) and *The Moral Judgment of the Child* (New York: Free Press, 1965); B. Bloom, ed., *Taxonomy of Educational Objectives, Handbook I: Cog-*

nitive Domain (New York: David McKay, 1956); D. Krathwohl, B. Bloom, and B. Masia, *Taxonomy of Educational Objectives, Handbook II: Affective Domain* (New York: David McKay, 1964); T. Lickona, *Educating for Character* (New York: Bantam Books, 1991); T. Lickona, ed., *Charmichael's Manual of Child Psychology* (New York: Wiley, 1970); T. Lickona, ed., *Moral Development and Behavior* (New York: Holt, Rinehart & Winston, 1976); J. Dunlap, "Ethical Reasoning About Animal Treatment and Its Relationship to Moral Development" (Ph.D. dissertation, Yale University, 1987).

6. These results, detailed in the previously noted U.S. Fish and Wildlife Service report on children's attitudes, are summarized in "Attitudes Toward Animals: Age-Related Development Among Children," *Journal of Environmental Education.* 16(1985):26–39.
7. For two highly informative and insightful recent publications regarding relationships among children and nature in America see G. P. Nabhan and S. Trimble, *The Geography of Childhood: Why Children Need Wild Places* (Boston: Beacon Pess, 1994); R. M. Pyle, *The Thunder Tree: Lessons from an Urban Wildland* (Boston: Houghton Mifflin, 1993). A classic essay worth reading is E. Cobb's *The Ecology of Imagination in Childhood* (New York: Columbia University Press, 1959).
8. This research is summarized in "Attitudes, Knowledge, and Behaviors Toward Wildlife as Affected by Gender," *Wildlife Society Bulletin* 15(1987): 363–371.
9. C. Gilligan, *In a Different Voice* (Cambridge: Harvard University Press, 1982).
10. Significant demographic variations among natural resource professionals have been described in various publications by J. Kennedy of Utah State University.
11. Trends in attitudes toward environmental protection and natural resource conservation have been monitored for more than two decades by R. Dunlap of the University of Washington and R. C. Mitchell of Worcester Polytechnic Institute.
12. Dunlap, Mitchell, and others have documented substantial urban/rural variations in resource use and perception. Other relevant publications include G. C. Gray, *Wildlife and People: The Human Dimensions of Wildlife Ecology* (Urbana: University of Illinois Press, 1993); M. Duda, *Americans and Wildlife Diversity* (Washington, D.C.: International Association of Fish and Wildlife Agencies, 1994).
13. Useful sources of information concerning environmental value differences among African-Americans and European-Americans include E. J. Dolan,

"Black Americans' Attitudes Toward Wildlife," *Journal Environmental Education* 20(1988):17–21; J. A. Caron, "Environmental Perspectives of Blacks: Acceptance of the 'New Environmental Paradigm,'" *Journal Environmental Education* 20(1989):21–26; D. E. Taylor, "Blacks and the Environment: Toward an Explanation of the Concern and Action Gap Between Blacks and Whites," *Environmental and Behavior* 21(1989):175–205; and two of my publications: "Urban American Perceptions of Animals and the Natural Environment," *Urban Ecology* 8(1984):209–228, and *Children's Attitudes, Knowledge, and Behaviors Toward Animals* (with M. Westervelt).

14. A detailed description of these findings can be found in my publication, *Urban American Perceptions and Uses of Animals and the Natural Environment.*
15. E. Cleaver, "The Land Question and Black Liberation," in R. Scheer, ed., *Post Prison Writings and Speeches* (New York: Random House, 1969), p. 58. See also J. W. Meeker, "Red, White and Black in the National Parks," *North American Review* 258(1973):3–7.
16. See, for example, P. H. Kahn, Jr., and B. Friedman, "Environmental Views and Values of Children in an Inner-City Black Community," *Child Development* (in press).

CHAPTER 4: ACTIVITIES

1. R. Pyle, *The Thunder Tree,* offers important insights, as does G.P. Nabhan and S. St. Antoine's essay, "The Loss of Floral and Faunal Story: The Extinction of Experience," in *The Biophilia Hypothesis.*
2. See, for example, R. B. Lee and I. DeVore, eds., *Man the Hunter* (Chicago: Aldine, 1968); S. L. Washburn, ed., *The Social Life of Early Man* (Chicago: Aldine, 1961); R. Ardrey, *The Hunting Hypothesis* (New York: Atheneum, 1976).
3. This provocative suggestion has been made by, among others, P. Shepard in *The Tender Carnivore and the Sacred Game* (New York: Scribner's, 1973).
4. A detailed description of these trends can be found in S. J. Bissell and M. D. Duda, *Factors Related to Hunting and Fishing Participation in the United States, Phase II: Hunting Focus Groups* (Harrisonburg, Va.: Responsive Management, 1993).
5. J. Ortega y Gasset, *Meditations on Hunting,* p. 6.
6. Among the many historical analyses of hunting in America, worthwhile references include G. Reiger, *American Sportsmen and the Origins of Conservation* (New York: Winchester Press, 1975); P. Matthiessen, *Wildlife in America;* and J. Tober, *Who Owns the Wildlife?*

7. P. Matthiessen, *Wildlife in America,* p. 192, taken from John James Audubon's *The Ornithological Biography* (Edinburgh, 1931–1939), 5 vol.
8. This development is described by J. Tober in *Who Owns the Wildlife?* and T. Lund, *American Wildlife Law* (Berkeley: University of California Press, 1980).
9. Data on hunting participation and trends are thoroughly summarized by M. D. Duda, *Factors Related to Hunting and Fishing Participation in the United States.* The Fish and Wildlife Service has also published reports every five years since 1955 on hunting and fishing participation rates; the last one was *National Survey of Fishing, Hunting, and Wildlife-Associated Recreation* (Washington, D.C.: Fish and Wildlife Service, 1993).
10. The importance of cultural context in hunting participation has been described by J. Applegate in a number of publications, most recently "Patterns of Early Desertion Among New Jersey Hunters," *Wildlife Society Bulletin* 17(1991):476–481. Other useful examinations of the cultural context of hunting include T. Alterr, *The Best of All Breathing: Hunting as a Mode of Environmental Perception in American Literature* (Ann Arbor: University Microfilms, 1976); J. Dizard, *Going Wild: Hunting, Animal Rights, and the Contested Meaning of Nature* (Amherst: University of Massachusetts Press, 1994); J. Swan, *In Defense of Hunting* (San Francisco: Harper, 1995).
11. M. D. Duda's monograph cited in note 9 provides important data on antihunting trends.
12. Useful references on the history of the humane movement, antihunting sentiment, and animal rights perspectives can be found in G. Carson, *Men, Beasts and Gods* (New York: Scribner's, 1972); K. Thomas, *Man and the Natural World* (New York: Pantheon Books, 1983); T. Regan, *All That Dwell Therein: Animal Rights and Environmental Ethics* (Berkeley: University of California Press, 1982); P. Singer, *Animal Liberation: A New Ethics for Our Treatment of Animals* (New York: New York Review, 1975); P. Taylor, *Respect for Nature* (Princeton: Princeton University Press, 1986); R. Morris and M. Fox, eds., *On the Fifth Day: Animal Rights and Human Ethics* (Washington, D.C.: Acropolis Books, 1978).
13. See S. Flader's biography, *Thinking Like a Mountain;* J. B. Callicott, ed., *Companion to A Sand County Almanac* (Madison: University of Wisconsin Press, 1987); the chapter on Aldo Leopold in M. Oelschlaeger, *The Idea of Wilderness.*
14. P. Shepard, A Theory of the Value of Hunting, *Transactions of the North American Wildlife and Natural Resources Conference* 24(1959):504–512.
15. J. Ortega y Gasset, *Meditations on Hunting,* p. 119.

16. J. Madson and E. Kozicky, "The Hunting Ethic," *Rod and Gun* 66(1964):12.
17. J. Ortega y Gasset, *Meditations on Hunting,* pp. 110–111.
18. C. H. D. Clarke, "Autumn Thoughts of a Hunter," *Journal of Wildlife Management* 22(1958):420–426.
19. V. Bourjailly, *The Unnatural Enemy* (New York: Dial Press, 1963), p. 15.
20. An insightful assessment of views for and against hunting can be found in W. W. Shaw, "Sociological and Psychological Determinants of Attitudes Toward Hunting" (Ph. D. dissertation, University of Michigan, 1974).
21. J. W. Krutch, "The Sportsmen or the Predator? A Damnable Pleasure," *Saturday Review,* August 19, 1957, p. 9.
22. Albert Schweitzer, "The Ethic of Reverence for Life," in T. Regan and P. Singer, eds., *Animal Rights and Human Obligations* (Englewood Cliffs: Prentice-Hall, 1976).
23. From an unpublished manuscript by P. Breer, "Can Hunting Be Justified?"
24. Many of the previously cited references include strong attacks against fur trapping. Additionally informative reading includes G. Nilsson, *Facts About Furs* (Washington, D.C.: Animal Welfare Institute, 1980); C. Amory, *Man Kind?* (New York: Harper & Row, 1974).
25. See, for example, C. Safina, "Where Have All the Fishes Gone?," *Issues of Science and Technology* 10(1994):37; P. Lopes, "A Critical Review of the Individual Quota as a Device in Fisheries Management," *Land Economics* 62(1986):278–288; J. Baum, "Nets Across the Strait," *Far Eastern Economic Review* 156(1993):22–28; G. Bush, "Statement on Signing the High Seas Driftnet Fisheries Enforcement Act," *Weekly Compilation of Presidential Remarks* 28(1992):2281; L. Craft, "$20,000 for One Fish? Bluefin Tuna May Be Worth Too Much for Their Own Good," *International Wildlife* 24(1991):18–21; M. Garrett, "Fishing: Pursuit of the Bluefins Meets Reality," *Sea Frontiers* 39(1993):18–20; D. MacKenzie, "Too Little, Too Late, to Save Atlantic Bluefin," New Scientist 140(1993):11.
26. A useful review of trends and participation rates regarding wildlife observation can be found in M. D. Duda and K. C. Young, *Americans and Wildlife Diversity* (Harrisonburg: Responsive Management, 1994); the previously cited U.S. Fish and Wildlife five-year reports; W. W. Shaw and W. R. Mangun, *Nonconsumptive Use of Wildlife in the United States,* (U.S. Fish and Wildlife Service Resource Publication 154, 1986); and a publication of mine, "Birdwatching in American Society," *Leisure Sciences* 7(1986): 343–360.
27. Useful data on trends in ecotourism can be found in T. Whelan, ed., *Nature Tourism* (Washington, D.C.: Island Press, 1991); T. W. Swanson and

E. B. Barbier, eds., *Economics for the Wilds* (Washington, D.C.: Island Press, 1992).

28. Useful statistics can be found in E. Hoyt, "Whalewatching Around the World: A Report on Its Value, Extent, and Prospects," *Journal of the Whale Conservation Society* 7 (1994); my paper with V. Scheffer, "The Changing Place of Marine Mammals in American Thought," in J. R. Twiss, Jr. and J. E. Reynolds III, eds., *Marine Animals* (Washington, D.C.: Smithsonian Press, 1996).
29. See M. Duda and K. Young, *Americans and Wildlife Diversity.*
30. Some of this information is included in my publication, "Attitudes, Knowledge and Behavior Toward Wildlife Among the Industrial Superpowers," *Journal Social Issues* 42(1993):53–69.
31. J. Kastner, *A World of Watchers* (New York: Knopf, 1986).
32. For a useful examination of converging historical factors in the evolution of the American conservation movement see T. Cart, *The Struggle for Wildlife Protection in the United States* (Ann Arbor: University Microfilms, 1971).
33. One of the finest essays written on this topic is B. Wilkes, "The Myth of the Nonconsumptive User," *Canadian Field-Naturalist* 91(1977):343–349.
34. Some statistics on zoo visitation can be found in J. Dunlap and S. Kellert, "Zoos and Zoological Parks," in W. T. Reich, ed., *Encyclopedia of Bioethics* (New York: Macmillan, 1994); J. Cherfas, *Zoo 2000: A Look Behind the Bars* (London: British Broadcasting System, 1984); A. Nelson, "Going Wild," *American Demographics* 50(1990):34–37.
35. This element of authenticity and direct enjoyment of animals at zoos is described by B. Birney, *A Comparative Study of Children's Perceptions and Knowledge of Wildlife as They Relate to Field Trip Experiences at the Los Angeles County Museum of Natural History and the Los Angeles Zoo* (Ann Arbor: University Microfilms, 1986).
36. For useful references regarding the educational and research benefits of the zoo see J. C. Coe, "Design and Perception: Making the Zoo Experience Real," *Zoo Biology* 4(1985):197–208; W. G. Conway, "Zoo and Aquarium Philosophy," in K. Sausman, ed., *Zoological Park and Aquarium Fundamentals* (Wheeling, W. Va.: American Association of Zoological Parks and Aquariums, 1982); M. Robinson, "Zoos Today and Tomorrow," *Anthrozöos* 2(1989):10–14; D. Chiszar et al., "For Zoos," *Psychological Record* 40(1990):3–13; J. Luoma, *A Crowded Ark: The Role of Zoos in Wildlife Conservation* (Boston: Houghton Mifflin, 1987); C. Tudge, *Last Animals at the Zoo* (London: Hutchinson Radius, 1991).

37. Some illustrative criticisms of the zoo can be found in T. Regan, *The Case for Animal Rights* (Berkeley: University of California Press, 1983); M. Clifton, "Chucking Zoo Animals Overboard: How and Why Noah Culls the Ark," *Animals Agenda* 8(1988):14–22, 53–54; M. Fox, *Inhumane Society: The American Way of Exploiting Animals* (New York: St. Martin's Press, 1990); D. Jamieson, "Against Zoos," in P. Singer, ed., *In Defense of Animals* (Oxford: Basil Blackwell, 1985).
38. Isak Dinesen, *Out of Africa* (New York: Vintage Books, 1989), pp. 287–288.
39. These sentiments are elaborated in two of my publications: "The Educational Potential of the Zoo and Its Visitor," *Philadelphia Zoo Review* 3(1987):7–13; *Informal Learning at the Zoo: A Study of Knowledge and Attitude Impacts* (Philadelphia: Zoological Society of Philadelphia) (with Julie Dunlap).
40. Some historical perspective on the zoo can be found in H. Heideger, *Man and Animal in the Zoo* (New York: Delacorte Press, 1969) and *Wild Animals in Captivity* (New York: Dover, 1964); J. Klaits and B. Klaits, eds., *Animals and Man in Historical Perspective* (New York: Harper & Row, 1974); J. Luoma, *A Crowded Ark;* H. Ritvo, *The Animal Estate: The English and Other Creatures in the Victorian Age* (Cambridge: Harvard University Press, 1987).
41. W. Donaldson, "Welcome," in P. Chambers, ed., *Conference on Informal Learning* (Philadelphia: Philadelphia Zoological Garden, 1987), p. 3.
42. For studies of zoo visitors see S. Bitgood and A. Benefield, *Visitor Behavior: A Comparison Across Zoos,* Technical Report 86-20 (Jacksonville: Jacksonville State University, 1987); C. A. Hill, "An Analysis of the Zoo Visitor," *Education* 11(1971):158–165; L. LaResche, *What People Do at the Zoo* (Baltimore: Baltimore Zoological Society, 1974); D. L. Marcellini and T. A. Jensen, "Visitor Behavior in the National Zoo's Reptile House," in Chambers, *Conference on Informal Learning;* S. B. Rosenfeld, *Informal Learning in Zoos* (Ann Arbor: University Microfilms, 1980); S. Swensen, "Comparative Study of Zoo Visitors at Different Types of Facilities" (unpublished manuscript, Yale University School of Forestry and Environmental Studies, 1982); S. Kellert and J. Dunlap, *Informal Learning at the Zoo.*
43. From S. Swensen, "Comparative Study of Zoo Visitors."
44. In addition to my previously noted publications, see also my "Zoological Parks in American Society," *Proceedings of the 1979 American Association of Zoological Parks and Aquariums* (Wheeling: American Association of Zoological Parks and Aquariums, 1982).
45. E. D. Asper et al., "Marine Mammals in Zoos, Aquaria, and Marine Zoological Parks in North America," *International Zoo Yearbook* 27(1988):287–294.

46. Data on marine mammals in captivity can be found in two of my publications: *Canadian Perceptions of Marine Mammal Conservation and Management in the Northwest Atlantic,* Technical Report 91-04 (Guelph, Ontario: International Marine Mammal Association, 1991); *The Changing Place of Marine Mammals in American Thought* (with Victor Scheffer).
47. Many of the ideas and supporting information for this section derive from my publication, "Wildlife and Film: A Relationship in Search of Understanding," *BKTS Journal* 10(1986):38–63 (with C. McConnell, S. Hamby, and V. Dompka). Informative studies of wildlife film and television include R. Fortner, "Influence of an Environmental Documentary on Knowledge and Attitudes" (Columbus: Ohio State University, School of Natural Resources, 1983); C. Parsons, *True to Nature* (Cambridge: Patrick Stephens, 1982); C. Underwood, *Images of Nature on Television* (Downsview, Ontario: York University, 1982).
48. A. Felthous and I published a series of papers on this research, as well as a literature review of the subject. These publications include: "Childhood Cruelty to Animals and Later Aggression Against People: A Review," *American Journal of Psychiatry* 144(1987):710–717; "Psychosocial Aspects of Selecting Animal Species for Physical Abuse," *Journal of Forensic Science* 32(1987):1713–1723; "Violence Against Animals and People," *Bulletin of the American Academy Psychiatry and Law* 14(1986):55–69; "Childhood Cruelty Toward Animals Among Criminals and Noncriminals," *Human Relations* 38(1985):1113–1129.
49. K. Thomas, *Man and the Natural World,* p. 119.
50. R. W. Ten Bensel, "Historical Perspectives of Human Values for Animals and Vulnerable People," *Proceedings of the Conference on the Human-Animal Bond* (Minneapolis: University of Minnesota, 1983).
51. Popular associations of abuse of animals and violence toward people can be found in R. Lockwood and G. Hodge, "The Tangled Web of Animal Abuse," *Humane Society News* 31(1986):10–15.
52. The literature review cited earlier, "Childhood Cruelty to Animals and Later Aggression Against People: A Review," covers much of the scientific research on this subject and assesses their varying strengths and weaknesses. For early studies linking a triad of animal abuse, fire-setting, and bedwetting see D. S. Hellman and N. Blackman, "Enuresis, Firesetting and Cruelty to Animals," *American Journal Psychiatry* 122(1966):1431–1435; D. E. Wax and V. G. Haddox, "Enuresis, Firesetting, and Animal Cruelty in Male Adolescent Delinquents," *Journal of Psychiatric Law* 2(1974):45–72; A. R. Felthous and H. Bernard, "Firesetting and Cruelty to Animals," *Forensic Science*

29(1979):240–246. Another relevant publication is L. DeViney, J. Dickert, and R. Lockwood, "Care of Pets Within Child Abusing Families," *International Journal for the Study of Animal Problems* 4(1983):321–336.

CHAPTER 5: SPECIES

1. Additional information on these results can be obtained in my publications: *Knowledge, Affection, and Basic Attitudes Toward Animals in American Society* (with J. Berry) and "Public Perceptions of Predators, Particularly the Wolf and Coyote," *Biological Conservation* 31(1985):167–189.
2. For informative publications on wolf history, folklore, and biology see L. D. Mech, *The Wolf: The Ecology and Behavior of an Endangered Species* (New York: Doubleday, 1981); B. Lopez, *Of Wolves and Men* (New York: Scribner's, 1978); P. Matthiessen, *Wildlife in America;* R. T. Dunlap, *Saving America's Wildlife* (Princeton: Princeton University Press, 1988); S. Young, *The Wolf in American History* (Caldwell: Caxton, 1946); E. Zimen, *The Wolf: Species in Danger* (New York: Delacorte Press, 1981); J. Nee, ed., *Wolf! Wolves in American Culture* (Ashland: Northwood Press, 1988); B. Ferris (filmmaker) and B. Weide (producer), *The Wolf: Real or Imagined?* (Missoula: Lone Wolf Productions and Montanans for Quality Television, 1992).
3. Informative explorations of American settlers' perceptions of wilderness include R. Nash, *Wilderness and the American Mind,* and M. Oelschlaeger, *The Idea of Wilderness.* The quote from John Adams is taken from B. Ferris and B. Weide, *The Wolf: Real or Imagined?*
4. S. Young, *The Wolf in American History,* p. 46.
5. B. Lopez, *Of Wolves and Men,* pp. 180–181, 184, and 148.
6. P. Matthiessen, *Wildlife in America,* p. 167.
7. B. Lopez, *Of Wolves and Men,* p. 137.
8. K. Dunlap, *Saving America's Wildlife,* p. 26.
9. B. Lopez, *Of Wolves and Men,* p. 163.
10. See S. Flader, *Thinking Like a Mountain.*
11. A. Leopold, *Sand County Almanac,* p. 67.
12. The wolf has been positively depicted in many popular books and film specials—for example, F. Mowat, *Never Cry Wolf* (New York: Bantam Books, 1963). For a more scientific but widely read book see A. Murie, *The Wolves of Mt. McKinley,* Fauna Series, no. 5 (Washington, D.C.: U.S. National Park Service, 1944). At least four environmental organizations currently focus considerable resources on wolf conservation and recovery: Defenders of Wildlife, The Wolf Fund, The International Wolf Center, and The Wolf Conservation and Education Fund. In 1993, ABC-TV aired an hour-long

prizewinning documentary on the wolf, *Wolf: Return of a Legend,* filmed and produced by J. Dutcher.

13. Variations in public perceptions of wolves can be found in a number of my publications: "Public Perceptions of Predators, Particularly the Wolf and Coyote"; "The Public and the Timber Wolf in Minnesota," *Transactions of the North American Wildlife and Natural Resources Conference* 51(1986):193–200; "Public Views of Wolf Restoration in Michigan," *Transactions of the North American Wildlife and Natural Resources Conference* 56(1991):152–161; *Restoration of Wolves in North America,* Technical Review 91–1 (Washington, D.C.: Wildlife Society, 1991) (with J. Peek et al.); "Perceptions of Wolves, Mountain Lions, and Grizzly Bears in North America," *Conservation Biology* (forthcoming) (with M. Black, C. Rush, and A. Bath). Other relevant publications include A. J. Bath, "Public Attitudes in Wyoming, Montana, and Idaho Toward Wolf Restoration in Yellowstone National Park," *Transactions of the North American Wildlife and Natural Resources Conference* 56(1991):91–95; C. Buys, "Predator Control and Ranchers' Attitudes," *Environment and Behaving* 7(1975):81–89; R. A. Hook and W. L. Robinson, "Attitudes of Michigan Citizens Toward Predators," in F. H. Harrington and P. C. Paquet, eds., *Wolves of the World* (Park Ridge: Noyes, 1982); L. Llewellyn, "Who Speaks for the Timber Wolf?" *Transactions of the North American Wildlife and Natural Resources Conference* 43(1978):442–452; D. A. McNaught, "Wolves in Yellowstone Park? Park Visitors Respond," *Wildlife Society Bulletin* 15(1987):518–521; P. Tucker and D. Pletscher, "Attitudes of Hunters and Residents Toward Wolves in Northwestern Montana," *Wildlife Society Bulletin* 17(1989):509–514; R. Johnson, "On the Spoor of the 'Big Bad Wolf,'" *Journal Environmental Education* 6(1974):37–39; F. H. Wagner, *Predator Control and the Sheep Industry* (Claremont: Regina Books, 1988).
14. S. Devlin, "Wolf Provokes Inadvertent Howlers," *High Country News,* September 19, 1994, p. 3.
15. B. Lopez, *Of Wolves and Men,* p. 145.
16. Reflecting this perspective, G. Grosvenor commented in 1976: "The whale has become a symbol for a way of thinking about our planet" (editorial, *National Geographic* 150:721). Relevant popular books and other mass media in the 1960s and 1970s reflecting significant shifts in attitudes toward whales include: S. McVay, "The Last of the Great Whales," *Scientific American* 215(1966):13–21; J. C. Lilly, *The Mind of the Dolphin* (Garden City: Doubleday, 1967); V. Scheffer, *The Year of the Whale* (New York: Schribner's, 1969); G. Small, *The Blue Whale* (New York: Columbia University Press,

1971); F. Mowat, *A Whale for the Killing* (New York: Little, Brown, 1962); J. McIntyre, ed., *Mind in the Waters* (New York: Scribner's and Sierra Club Books, 1974); the recordings of humpback whales by R. Payne, *Songs of the Humpback Whale* (Los Angeles: New York Zoological Society and Communications/Research/Machines, 1970), and their incorporation into the music of Judy Collins (*Whales and Nightingales,* Elektra 76–7638501, 1970), George Crumb (*Vox Balaenae: Voice of the Whale,* Columbia Records, M-32739, 1974), and Paul Winter (*Callings,* Living Music Records, 1980). These and other references were collected by V. Scheffer and can be found in our paper "The Changing Place of Marine Mammals in American Thought."

17. Historical descriptions of whaling can be found in J. Scharff, "The International Management of Whales, Dolphins, and Porpoises: An Interdisciplinary Assessment," *Ecology Law Quarterly* 6(1977):243–352; S. F. Harmer, "History of Whaling," *Proceedings of the Linnaean Society* 140(1928):51–95; J. L. McHugh, "The Role and History of the International Whaling Commission," in W. E. Schevill, ed., *The Whale Problem* (Cambridge: Harvard University Press, 1974); R. Burton, *The Life and Death of Whales* (London: Andre Deutsch, 1973); Friends of the Earth, *The Whale Manual* (San Francisco: Friends of the Earth, 1978); K. Norris, "Marine Mammals and Man," in H. P. Brokaw, ed., *Wildlife and America* (Washington, D.C.: Council on Environmental Quality, 1978); S. Lyster, *International Wildlife Law* (Cambridge, England: Grotius, 1985).
18. This quote and many other statistics on whaling and whale management are drawn from the excellent paper by K. Norris, "Marine Mammals and Man," p. 320. The D. Ehrenfeld statistic appears in *Biological Conservation* (New York: Holt, Rinehart & Winston, 1970). The statistics on current numbers of endangered and threatened marine mammals are obtained from the U.S. Marine Mammal Commission.
19. Considerable material for this chapter is drawn from my paper, "Marine Mammals, Endangered Species, and Intergovernmental Relations," in M. Silva, ed., *Ocean Resources and U.S. Intergovernmental Relations in the 1980s* (Boulder: Westview Press, 1986); P. Matthiessen, *Wildlife in America,* p. 107.
20. G. Hardin, "The Tragedy of the Commons," *Science* 163(1954):1243. Important revisions of Hardin's thesis can be found in D. Feeny et al., "The Tragedy of the Commons: Twenty-Two Years Later," *Human Ecology* 18(1990):1–19; F. Berkes, "Fishermen and the Tragedy of the Commons," *Environmental Conservation* 12(1985):199–206.

21. See my paper with V. Scheffer, "The Changing Place of Marine Mammals in American Thought."
22. For relevant sources of information on trends in whalewatching see J. Hoyt, *Whalewatching Around the World;* S. Kaza, "Recreational Whalewatching in California," *Whalewatcher* 16(1982):6–8; J. E. Kelly, "The Value of Whale-Watching, *Transactions of the Global Conference on the Non-consumptive Utilization of Cetaceans,* (1983).
23. See R. L. Wallace, comp., *The Marine Mammal Commission Compendium of Selected Treaties, International Agreements, and Other Relevant Documents on Marine Resources, Wildlife, and the Environment* (Washington, D.C.: Marine Mammal Commission, 1994); M. Bean, *The Evolution of National Wildlife Law.*
24. See my publications: "Canadian Attitudes Toward Marine Mammal Management and Conservation in the Northwest Atlantic"; "Canadian Perceptions of Commercial Fisheries and Marine Mammal Management," *Anthrozöos* 8(1995):20–30 (with J. Gibbs and T. Wohlgenant); "International Attitudes to Whales, Whaling, and the Use of Whale Products," in M. Freeman and U. Kreuter, eds., *Elephants and Whales: Resources for Whom?* (Reading, UK: Gordon & Breach, 1995) (with M. Freeman).
25. See E. O. Wilson, *The Diversity of Life;* B. Groombridge, *Global Biodiversity.*
26. Much of the information for this section comes from my publication, "Values and Perceptions of Invertebrates," *Conservation Biology* 7(1993): 845–855, which offers a great many references on the ecological, utilitarian, commercial, and other values of invertebrates. For other useful sources see the references in note 25 and J. L. Cloudsley, *Insects and History* (New York: St. Martin's Press, 1976); C. J. Krebs, *Ecology* (New York: Harper & Row, 1978); C. Lindberg, "The Economic Value of Insects," *Traffic Bulletin* 10(1993):32–36; R. M. May, *Stability and Complexity in Model Ecosystems* (Princeton: Princeton University Press, 1973); D. Pimentel, *Insects, Science and Society* (New York: Academic Press, 1975); T. R. E. Southwood, "Entomology and Mankind," *American Science* 65(1992):30–39; J. Adams, ed., *Insect Potpourri: Adventures in Entomology* (Gainesville: Sandhill Crane Press, 1992).
27. See E. O. Wilson, *The Diversity of Life;* U.S. Fish and Wildlife Service, *Official List of Endangered and Threatened Species* (Washington, D.C.: U.S. Department of Interior, Office of Endangered Species, 1994).
28. Much of this section derives from my paper "Values and Perceptions of Invertebrates."

29. This comment appeared in a letter I received in response to a December 21, 1993, *New York Times* newspaper article, "Bugs Keep Planet Livable Yet Get No Respect," written by W. K. Stevens, based largely on my *Conservation Biology* paper on people and invertebrates.
30. I have benefited greatly from the wisdom and insight of James Hillman's audiotape, *Going Bugs* (Gracie Station: Spring Audio, 1991). Other related insights can be found in T. Moore and J. Hillman, eds., *A Blue Fire: Selected Writings of James Hillman* (New York: Harper & Row, 1989).
31. All the Hillman quotes are from *Going Bugs.*

CHAPTER 6: CULTURE

1. This issue was the focus of a recent book of informative and engaging essays: M. E. Soulé and G. Lease, eds., *Reinventing Nature? Responses to Postmodern Deconstruction* (Washington, D.C.: Island Press, 1994). My chapter in that book is entitled "Concepts of Nature East and West," and many of these ideas were presented there.
2. L. White, "The Historic Roots of Our Ecologic Crisis," *Science* 176(1967): 1203–1207. Other important insights can be found in K. Thomas, *Man and the Natural World;* M. Oelschlager, *The Idea of Wilderness;* J. Ortega y Gasset, *Man and Crisis* (New York: Norton, 1962); W. Leiss, *The Domination of Nature* (New York: Braziller, 1972); J. Passmore, *Man's Responsibility for Nature: Ecological Problems and Western Traditions* (New York: Scribner's, 1974). The following L. White quotes are from "The Historic Roots of Our Ecologic Crisis, pp. 1205–1206.
3. K. Thomas, *Man and the Natural World,* p. 237.
4. For informative references on Eastern conceptions of nature see J. B. Callicott and R. T. Ames, eds., *Nature in Asian Traditions of Thought* (Albany: State University of New York, 1989); N. Nash, ed., *Tree of Life: Buddhism and Protection of Nature* (Hong Kong: Buddhism Protection of Nature Project, 1987); M. Anesaki, *Art, Life, and Nature in Japan* (Rutland: Tuttle, 1932); K. Higuchi, *Nature and the Japanese* (Tokyo: Kodansha International, 1979); R. Nash, *The Rights of Nature: A History of Environmental Ethics* (Madison: University of Wisconsin Press, 1989); Y. Saito, *The Aesthetic Appreciation of Nature;* D. T. Suzuki, *Zen and Japanese Culture* (Princeton: Princeton University Press, 1973).
5. H. Watanabe, "The Conception of Nature in Japanese Culture," *Science* 183(1973):280. All the Watanabe quotes in this chapter are from this essay.
6. In addition to the references cited in note 4, the following publications are cited here and elsewhere in this chapter: F. Mariani, *Japan: Patterns of Con-*

tinuity (New York: Harper & Row, 1971), p. 19; Y. Murota, "Culture and the Environment in Japan," *Environmental Management* 9(1986):105; Y. Saito, *The Aesthetic Appreciation of Nature,* p. 47.

7. These environmentally damaging practices are noted, for example, in N. Kamei, "Clearing the Clouds of Doubt," *Mainstream* (1983):22–25; E. Linden, "Putting the Heat on Japan: Accused of Ravaging the World's Forests and Seas," *Time Magazine,* 1988; J. Luoma, "Japan Assailed on Animal Imports," *New York Times,* 1988; C. Moreby, "What Whaling Means to the Japanese," *New Scientist* 31(1982):661–663; M. Oyadomori, *Politics of National Parks in Japan* (Ann Arbor: University Microfilms, 1985); M. Sun, "Japan Prodded on the Environment," *Science* 241(1989):46.
8. See, for example, J. Tober, *Wildlife and the Public Interest: Nonprofit Organizations and Federal Wildlife Policy* (New York: Praeger, 1989).
9. This research has been summarized in my publication, "Japanese Perceptions of Wildlife," *Conservation Biology* 5(1991):297–308. Many of the statistics regarding Japanese biogeography, population, and biodiversity are taken from this paper. Additional sources of information include: Japanese Environment Agency, *The Natural Environment of Japan* (Tokyo, 1982); I. Yano, *Nippon: A Chartered Survey of Japan* (Tokyo: Kokuseisha, 1984); G. T. Trewartha, *Japan: A Geography* (Madison: University of Wisconsin Press, 1965).
10. A. Graphard, "Nature and Culture in Japan," in M. Tobias, ed., *Deep Ecology* (San Diego: Avant Books, 1985), p. 243. Another helpful essay, unpublished, is by R. Taylor, "On the Japanese Love of Nature" (1990).
11. Y. Saito, *The Aesthetic Appreciation of Nature,* p. 192. The following studies have also proved helpful in describing contemporary Japanese attitudes toward nature and its conservation: M. Imanaga et al., *International Comparison of Attitudes Toward Nature* (Kagoshima: Forestry Department, Kagoshima University); Japan Prime Minister's Office, *1986 and 1988 Public Opinion Surveys on Animal Conservation and Environmental Problems;* M. A. McKean, *Environmental Protest and Citizen Politics in Japan* (Berkeley: University of California Press, 1981); J. C. Pierce et al., *Public Knowledge and Environmental Politics in Japan and the United States* (Boulder: Westview Press, 1989); T. Shidei, *International Comparisons of Attitudes Toward Nature* (Tokyo: Japan Institute for Statistical Mathematics, 1983); M. Tsutsui, "Nature Viewed by Japanese and a New Approach to Its Management," *Transactions of the 22nd IUFRO World Congress;* F. K. Upham, "After Minimata: Current Prospects and Problems in Japanese Environmental Litigation," *Ecology Law Quarterly* 8(1979):213–268.

12. Personal communication, H. Watanabe.
13. These data are taken from my article, "Attitudes, Knowledge, and Behavior Toward Wildlife Among the Industrial Superpowers: United States, Japan, and Germany," *Journal Social Issues* 49(1993):53–69.
14. H. Watanabe, "The Conception of Nature in Japanese Culture," p. 281; B. Callicott and R. Ames, *Nature in Asian Traditions of Thought,* p. 280.
15. W. Schulz, "Einstellung zur Nature, eine empirische Untersuchung" (Ph. D. dissertation, University of Munich, 1985); "Attitudes Toward Wildlife in West Germany," in D. Decker and G. Goff, eds., *Valuing Wildlife* (Boulder: Westview Press, 1986). Much of the information on Germany has been taken from my publication cited in note 13 on attitudes toward nature and wildlife in Germany, United States, and Japan.
16. This quote is from an unpublished paper by A. Nelson, "History of German Forestry." Other helpful publications include R. Plochman, *Forestry in the Federal Republic of Germany* (St Paul: L & M Hill Foundation, 1968); M. Wolfe, "The History of German Game Administration," *Forest History* 14(1970):6–16; A. Leopold, *Game Management* (New York: Macmillan, 1933) and "Deer and Dauerwald in Germany," *Journal of Forestry* 34(1936): 366–375.
17. Much of the information on Botswana is taken from the book by R. Mordi, *Attitudes Toward Wildlife in Botswana* (New York: Garland, 1991).
18. Useful sources of information on Bushmen and hunter-gatherer perspectives of nature and animals include essays in *Proceedings of the Symposium on the Okavango Delta and Its Future Utilization* (Gaborone: Botswana National Museum, 1976); R. Lee, "The !Kung Bushmen of Botswana," and other essays in M. Micchieri, ed., *Hunters and Gatherers Today* (New York: Holt, Rinehart & Winston, 1972); S. Marks, *The Imperial Lion: Human Dimensions of Wildlife Management in Central Africa* (Boulder: Westview Press, 1984); M. and D. Owens, *Cry of the Kalahari* (Boston: Houghton Mifflin, 1984); E. Hunn and N. Williams, eds., *Resource Managers: North American and Australian Hunter-Gatherers* (Boulder: Westview Press, 1982); R. Redfield, *The Primitive World and Its Transformations* (Ithaca: Cornell University Press, 1953); J. B. Callicott, "Traditional European and American Indian Attitudes Toward Nature: An Overview," *Environmental Ethics* 4(1982):293–318. The excellent insights and understanding of Richard Nelson have been particularly helpful, including: *Hunters of the Northern Ice* (Chicago: University of Chicago Press, 1969); *Make Prayers to the Raven* (Chicago: University of Chicago Press, 1983); and *The Island Within* (Chicago: University of Chicago Press, 1991).

19. From R. Nelson's important essay, "Searching for the Lost Arrow," in *The Biophilia Hypothesis*, pp. 212–215.
20. See, for example, B. Neitschmann, *Between Land and Water: The Subsistence Ecology of the Miskito Indians, Eastern Nicaragua* (New York: Seminar Press, 1973).

CHAPTER 7: ENDANGERED SPECIES

1. For a detailed description of the policy framework see my publications with T. Clark: "The Theory and Application of a Wildlife Policy Framework," in W. R. Mangun and S. S. Nagel, eds., *Public Policy and Wildlife Conservation* (New York: Greenwood Press, 1991); "Toward a Policy Paradigm of the Wildlife Sciences," *Renewable Resources Journal* 6(1988):7–16.
2. For a thorough delineation of the temporal dimension of the policy process see G. Brewer and P. DeLeon, *The Foundations of Policy Analysis* (Homewood: Dorsey Press, 1983).
3. A number of excellent legal and policy descriptions of the Endangered Species Act are available: M. Bean, *The Evolution of National Wildlife Law* (New York: Praeger, 1983); D. J. Rohlf, *The Endangered Species Act: A Guide to Its Protection and Implementation* (Stanford: Stanford Environmental Law Society, 1990); K. A. Kolm, ed., *Balancing on the Brink of Extinction: The Endangered Species Act and Lessons for the Future* (Washington, D.C.: Island Press, 1991); S. L. Yaffee, *Prohibitive Policy: Implementing the Federal Endangered Species Act* (Cambridge: MIT Press, 1982); O. A. Houck, "The Endangered Species Act and Its Implementation by the U.S. Departments of Interior and Commerce," *University of Colorado Law Review* 64(1993):278–369; T. Clark, R. Reading, and A. Clarke, *Endangered Species Recovery: Finding the Lessons, Improving the Process* (Washington, D.C.: Island Press, 1994); G. Coggins and I. Russel, "Beyond Shooting Snail Darters in Pork Barrels: Endangered Species Use in America," *Georgetown Law Journal* 70(1985): 1433–1525.
4. See, for example, J. Echeverria and R. Eby, eds., *Let the People Judge: Wise Use and the Private Property Rights Movement* (Washington, D.C.: Island Press, 1995); J. S. Burling, "Property Rights, Endangered Species, Wetlands, and Other Critters—Is It Against Nature to Pay for a Taking?" *Land and Water Law Review* 27(1992):309–362; I. M. Heyman, "Property Rights and the Endangered Species Act: A Renascent Assault on Land Use Regulation," *Pacific Law Journal* 25(1994):157–170; C. C. Mann and M. Plummer, *Noah's Choice: The Future of Endangered Species* (New York: Knopf, 1995); S. Shaheen, "The Endangered Species Act: Inadequate Species Protection in the

Wake of the Destruction of Private Property Rights," *Ohio State Law Journal* 55(1994):453–476.

5. See the essays in T. Clark et al., *Endangered Species Recovery;* K. Kolm, *Balancing on the Brink of Extinction;* O. Houck, *The Endangered Species Act and Its Implementation;* and my chapter, "A Sociological Perspective: Valuational, Socioeconomic, and Organizational Factors in Endangered Species Implementation," in T. Clark et al., *Endangered Species Recovery.*
6. For two useful government reports delineating aspects of the Endangered Species Act's implementation see General Accounting Office, *Endangered Species: Management Improvements Could Enhance Recovery Program* (Washington, D.C.: Government Printing Office, 1989); General Accounting Office, *Endangered Species Act: Types and Number of Implementing Actions* (Washington, D.C.: Government Printing Office, 1992). Other useful references include D. J. Rohlf, "Six Biological Reasons Why the Endangered Species Act Doesn't Work—And What to Do About It," *Conservation Biology* 5(1991):273–282; W. Reffalt, "The Endangered Species Lists: Chronicles of Extinction?," in K. Kolm, *Balancing on the Brink of Extinction.*
7. An informative overview of the snail darter dispute can be found in Z.J.B. Plater, "In the Wake of the Snail Darter: An Environmental Law Paradigm and Its Consequences," *Journal of Law Reform* 19(1986):805–862. Another publication of mine also includes information on this conflict: "Assessing Wildlife and Environmental Values in Cost-Benefit Analysis," Journal of Environmental Management 18(1984):353–363.
8. Information on the palila dispute can be found in M. Bean, *The Evolution of National Wildlife Law,* and G. W. Coggins and C. Wilkinson, *Federal Public Land and Resources Law* (Mineola: Foundation Press, 1987).
9. A very informative book on the birds of Hawaii, with information on the palila, is A. J. Berger, *Hawaiian Birdlife* (Honolulu: University of Hawaii Press, 1981). Two other relevant publications include C. Van Riper, J. M. Scott, and D. M. Woodside, "Distribution and Abundance Patterns of the Palila on Mauna Kea, Hawaii," *Auk* 95(1978):518–527; J. M. Scott et al., "Forest Bird Communities of the Hawaiian Islands: Their Dynamics, Ecology, and Conservation," *Studies in Avian Biology* 9(1986).
10. Such variations in public sentiment toward endangered species are detailed in two of my publications: "Socioeconomic Factors in Endangered Species Management," *Journal of Wildlife Management* 49(1985):528–536; "Social and Perceptual Factors in the Preservation of Animal Species," in B. G. Norton, ed., *The Preservation of Species: The Value of Biological Diversity* (Princeton: Princeton University Press, 1986).

11. Relevant information on the black-footed ferret and prairie dog can be found in R. P. Reading, *Toward an Endangered Species Reintroduction Paradigm: A Case Study of the Black-Footed Ferret* (Ann Arbor: University Microfilms, 1993); T. W. Clark, "Restoration of the Endangered Black-Footed Ferret: A 20-Year Overview," in M. L. Bowles and C. J. Whelan, eds., *Restoration of Endangered Species* (New York: Cambridge University Press, 1994); T. W. Clark, "Conservation Biology of the Black-Footed Ferret," *Wildlife Preservation Trust International Special Scientific Report,* no. 3(1989): R. P. Reading and T. W. Clark, "Black-Footed Ferret Annotated Bibliography, 1986–1990," *Montana BLM Wildlife Technology Bulletin* 3(1990):1–22.
12. For informative publications on livestock production and the western rangelands see P. O. Foss, *Politics and Grass: The Administration of Grazing on the Public Domain* (New York: Greenwood, 1960); P. J. Culhane, *Public Land Politics: Interest Group Influence on the Forest Service and the Bureau of Land Management* (Baltimore: Johns Hopkins University Press, 1981); C. F. Wilkinson, *Crossing the Next Meridian: Land, Water, and the Future of the West* (Washington, D.C.: Island Press, 1992); R. F. Noss and A. Y. Cooperrider, *Saving Nature's Legacy* (Washington, D.C.: Island Press, 1994).
13. This pattern has been revealed in a number of my studies: "Attitudes Toward a Proposed Reintroduction of Black-Footed Ferrets," *Conservation Biology* 7(1993):569–580 (with R. Reading); "Canadian Perceptions Toward Marine Mammal Management and Conservation in the Northwest Atlantic"; "The Public and the Timber Wolf in Minnesota"; "Public Attitudes Toward Mitigating Energy Development Impacts on Western Mineral Lands," *Proceedings in Issues and Technology in Management of Impacted Western Wildlife* (Boulder: Thorne Ecological Institute, 1984).
14. See R. Reading, *Toward an Endangered Species Reintroduction Paradigm,* and my publication with R. Reading, "Attitudes Toward a Proposed Reintroduction of Black-Footed Ferrets."
15. These issues are discussed in my publication, "A Sociological Perspective," and other essays in the book edited by T. Clark, R. Reading, and A. Clarke, *Endangered Species Recovery;* see also my contribution to a paper submitted for publication to *Ecological Applications:* M. Mangel et al., "Principles for the Conservation of Wild Living Resources."
16. This issue has been effectively considered in a number of publications by T. W. Clark, including his edited volume, *Endangered Species Recovery.* See also his "Creating and Using Knowledge for Species and Ecosystem Conservation: Science, Organizations, and Policy," *Perspectives in Biology and*

Medicine 36(1993):497–525; "Organization and Management of Endangered Species Programs," *Endangered Species Update* 8(1991):1–4 (with J. Cragun); "High Performance Teams in Wildlife Conservation: A Species Reintroduction and Recovery Example," *Environmental Management* 13(1989):663–670 (with R. Westrum); "Designing and Managing Successful Endangered Species Recovery Programs," *Environmental Management* 13(1989):159–170 (with R. Crete and J. Cada). Other useful publications include T. Clarke and D. McCool, *Staking Out the Terrain;* S. Yaffee, *Prohibitive Policy*; and S. Yaffee, *The Wisdom of the Spotted Owl: Policy Lessons for a New Century* (Washington, D.C.: Island Press, 1994).

17. T. Clarke and D. McCool, *Staking Out the Terrain,* p. 2.
18. See, for example, D. P. Warwick, *A Theory of Public Bureaucracies* (Cambridge: Harvard University Press, 1975).
19. R. F. Noss, "From Endangered Species to Biodiversity," in K. Kohm, *Balancing on the Brink of Extinction.*
20. From *The Best Nature Writing of Joseph Wood Krutch.*

CHAPTER 8: CONSERVING BIOLOGICAL DIVERSITY

1. Useful references on indirect habitat-related impacts on biodiversity include E. O. Wilson, *The Diversity of Life;* E. O. Wilson and R. Peters, *Biodiversity*; R. E. Grumbine, *Ghost Bears: Exploring the Biodiversity Crisis* (Washington, D.C.: Island Press, 1992); B. Groombridge, *Global Biodiversity;* P. M. Vitousek, P. R. Ehrlich, A. H. Ehrlich, and P. A. Matson, "Human Appropriation of the Products of Photosynthesis," *BioScience* 36(1991):368–373.
2. Useful summaries can be found in R. Noss and A. Cooperrider, *Saving Nature's Legacy;* E. A. Norse, *Ancient Forests of the Pacific Northwest* (Washington, D.C.: Island Press, 1990); M. L. Hunter, *Wildlife, Forests, Forestry* (Englewood Cliffs: Prentice-Hall, 1990); J. L. Holecheck, R. D. Pieper, and C. H. Herbel, *Range Management: Principles and Practices* (Englewood Cliffs: Prentice-Hall, 1989); G. V. Burger, "Agriculture and Wildlife," in H. Brokaw, *Wildlife and America;* A. Savory, *Holistic Resource Management* (Washington, D.C.: Island Press, 1988); W. Berry, *What Are People For?* (Berkeley: North Point, 1990).
3. S. Leopold, "Wildlife and Forest Practice," in H. Brokaw, *Wildlife and America,* p. 117.
4. A. Grussow, *A Sense of Place: The Artist and the Land* (1972), p. 27.
5. Many observations regarding biodiversity and people in the modern city are detailed in my essay "Environmental Values, the Coastal Context, and a

Sense of Place," in M. Sagoff, ed., *Place, Locality, and Community in Enclosed Coastal Seas* (Princeton: Princeton University Press, 1995). Other relevant readings include M. Sagoff, "Settling America or the Concept of Place in Environmental Ethics," *Journal of Energy, Natural Resources, and Environmental Law* 12(1992):351–418; D. Orr, "Love It or Lose It: The Coming Biophilia Revolution," in *The Biophilia Hypothesis;* R. Pyle, *The Thunder Tree.*

6. These two observations appear in E. O. Wilson, *Biophilia,* and A. Leopold, *Sand County Almanac.*
7. See J. Heerwagen and G. Orians, "Humans, Habitats, and Aesthetics," in *The Biophilia Hypothesis;* and see J. Heerwagen and G. Orians, "Adaptations to Windowlessness: A Study of the Use of Visual Decor in Windowed and Windowless Offices," *Environment and Behavior* 18(1986):623–639.
8. See, for example, B. Wilkes, "The Myth of the Nonconsumptive User"; T. Whelan, "Nature Tourism: Managing for the Environment"; R. L. Knight and K. J. Gutzwiller, eds., *Wildlife and Recreationists: Coexistence Through Management and Research* (Washington, D.C.: Island Press, 1995); W. E. Hammitt and D. N. Cole, *Wildland Recreation: Ecology and Management* (New York: Wiley, 1987); K. Lindberg and D. E. Hawkins, eds., *Ecotourism: A Guide for Planners and Managers* (North Bennington: Ecotourism Society, 1993).
9. See, for example, P. Matthiessen, *Wildlife in America;* R. Dunlap, *Saving America's Wildlife;* S. Fitzgerald, *International Wildlife Trade: Whose Business Is It?* (Washington, D.C.: World Wildlife Fund, 1989); G. Hemley, "International Wildlife Trade," in W. P. Chandler and L. Labate, eds., *Audubon Wildlife Report 1988/89* (New York: Academic Press, 1989); T. Inskipp and S. Wells, *International Trade in Wildlife* (London: International Institute for Environment and Development, 1979); G. Hemley, ed., *International Wildlife Trade: A CITES Sourcebook* (Washington, D.C.: Island Press, 1994).
10. For useful references debating the pros and cons of the contemporary wildlife trade see J. B. Thomsen and R. Luxmore, "Sustainable Utilization of Wildlife for Trade," and S. Foster, "Some Legal and Institutional Aspects of the Economic Utilization of Wildlife" (paper for Workshop 7, Sustainable Utilization of Wildlife, 18th IUCN General Assembly, Gland, Switzerland, 1990); V. Geist, "How Markets in Wildlife Meat and Parts, and the Sale of Hunting Privileges, Jeopardize Wildlife Conservation," *Conservation Biology* 2(1988):15–26; J. McNab, "Does Game Cropping Serve Conservation? Reexamination of the African Data," *Canadian Journal of Zoology;* T. W. Swanson and E. B. Barbier, *Economics for the Wilds* (Washington, D.C.:

Island Press, 1992); R. B. Martin, "Communal Area Management Plan for Indigenous Resources (Project Campfire)," in R.H.V. Bell and E. McShane-Caluzi, eds., *Conservation and Wildlife Management in Africa* (Washington, D.C.: U.S. Peace Corps, 1984); K. Muir, "The Potential Role of Indigenous Resources in the Economic Development of Arid Environments in Sub-Saharan Africa: The Case of Wildlife Utilization in Zimbabwe," *Society and Natural Resources Journal* 2(1988):307–318; R. B. Martin, "Integrating the Needs of Local People into the Design of Conservation and Wildlife Management" (unpublished paper, Yale University School of Forestry and Environmental Studies, 1991); R. B. Martin, J. R. Caldwell, and J. G. Barzdo, *African Elephants, CITES, and the Ivory Trade* (Lausanne: CITES Secretariat, 1986); M. Freeman and U. Kreuter, eds., *Elephants and Whales: Resources for Whom?* (London: Gordon and Breach, 1995).

11. For useful readings about the African elephant see I. and O. Douglas-Hamilton, *Battle for the Elephants* (New York: Viking Press, 1992); J. Hanks, *The Struggle for Survival* (New York: Mayflower Books, 1993); E. R. Ricciuti, "The Ivory Wars," *Animal Kingdom* 83(1985):6–58; I. Parker and M. Amin, *Ivory Crisis* (London: Chatto & Windus, 1983); C. Moss, *Elephant Memories: Thirteen Years in the Life of an Elephant Family* (New York: Morrow, 1988); R. M. Nowak and J. L. Pardiso, *Walker's Mammals of the World* (Baltimore: Johns Hopkins University Press, 1983); I. Douglas-Hamilton, "African Elephants: Population Trends and Their Causes," *Oryx* 21(1987):11–24; Environmental Investigation Agency, *Under Fire: Elephants on the Front Line* (London: Environmental Investigation Agency, 1992); H. T. Dublin and H. Jachmann, *The Impact of the Ivory Ban on Illegal Hunting of Elephants in Six Range States in Africa* (Gland, Switzerland: World-Wide Fund for Nature, 1992); D. H. Chadwick, *The Fate of the Elephant* (San Francisco: Sierra Club Books, 1992).
12. Many arguments presented here are detailed in my publications: "The Theory and Application of a Wildlife Policy Framework" (with T. Clark); "A Sociological Perspective: Valuational, Socioeconomic, and Organizational Factors"; "Principles for the Conservation of Wild Living Resources" (with M. Mangel et al.); "Managing for Biological and Sociological Diversity or 'Deja Vu, All Over Again,'" *Wildlife Society Bulletin* 23(1995).

CHAPTER 9: EDUCATION AND ETHICS

1. Persuasive articulations of this need can be found in A. Leopold, *Sand County Almanac;* E. O. Wilson, *Biophilia;* D. W. Orr, *Earth in Mind: On Ed-*

ucation, Environment, and the Human Prospect (Washington, D.C.: Island Press, 1994); B. Norton, *Why Preserve Natural Variety?* (Princeton: Princeton University Press, 1987).

2. For useful readings on the philosophy of environmental education see D. Orr, *Ecological Literacy* (Albany: State University of New York Press, 1992); L. Gigliotti, "Environmental Education: What Went Wrong?," *Journal of Environmental Education* 22(1991):9–12; H. Hungerford and T. Volk, "Changing Learner Behavior Through Environmental Education," *Journal of Environmental Education* 21(1990):8–21; J. Ramsey et al., "Environmental Education in the K–12 Curriculum," *Journal of Environmental Education* 23(1992):35–45; L. Iozzi, *Preparing for Tomorrow's World* (New Brunswick: Rutgers University Press, 1983); "Environmental Education and the Affective Domain—Parts One and Two," *Journal of Environmental Education* 20(3)(1989):3–9, 20(4)(1989):6–12; M. Caduto, "A Review of Environmental Values Education," *Journal of Environmental Education* 14(1983):13–21; H. Hungerford et al., "A Framework for Environmental Education Curriculum Planning and Development," in *Yearbook of Environmental Education; Project Wild and Aquatic Project Wild* (Boulder: Project Wild, 1992); *Project Learning Tree* (Washington, D.C.: American Forest Institute, 1977); G. W. Sharpe, ed., *Interpreting the Environment* (New York: Wiley, 1982); M. O'Connor, *Living Lightly on the Planet* (Milwaukee: Schlitz Audubon Center, National Audubon Society, 1992).
3. Useful discussions of the media's role in educating the public about the environment can be found in C. LaMay and E. Dennis, eds., *Media and the Environment* (Washington, D.C.: Island Press, 1993); E. Lambeth, *Committed Journalism* (Bloomington: Indiana University Press, 1992).
4. Useful readings on environmental ethics and biodiversity include A. Leopold, *Sand County Almanac;* E. O. Wilson, *Biophilia;* H. Rolston, *Philosophy Gone Wild;* P. Wenz, *Environmental Justice* (Albany: State University of New York, 1988); B. Norton, *Why Preserve Natural Variety;* J. B. Callicott, *In Defense of the Land Ethic: Essays in Environmental Philosophy* (Albany: State University of New York Press, 1989); M. Sagoff, "Zuckerman's Dilemma: A Plea for Environmental Ethics," *Hastings Center Report* 21(1991):32–41.
5. Some of the ideas in this section are developed in my publications: "The Biological Basis for Human Values of Nature"; "Mending the Broken Circle," in F. H. Bormann and S. R. Kellert, eds., *Ecology, Economics, Ethics: The Broken Circle* (New Haven: Yale University Press, 1991) (with F. H. Bormann); "Social and Psychological Dimensions of an Environmental Ethic,"

Proceedings of the International Conference on Outdoor Education (Washington, D.C.: Izaak Walton League, 1990).

6. E. O. Wilson, "Biophilia and the Conservation Ethic," in *The Biophilia Hypothesis,* p. 37.
7. R. Nelson, "Searching for the Lost Arrow," in *The Biophilia Hypothesis,* p. 223.
8. E. O. Wilson, *Biophilia,* p. 22.

INDEX

Aboriginal peoples, 14
Abusing animals, 92–98
Accountability for wildlife agencies, 203
Activities involving animals/nature, 64
  abusing animals, 92–98
  birding, 79–83
  commercial hunting, 77–79
  film and television, 91–93, 141
  hunting, 65–77
  zoos, 83–89
Adams, John, 104
Administration of wildlife management, 203–8
Aesthetic value of nature, 6, 14, 32
  affective learning, 212
  birding, 81
  black-footed ferrets, 176
  Botswana, 148–49
  conservation policy, 170–71
  defining, 38
  demographics on American, 44
  elephants, 197
  invertebrates, 128–29
  Japanese-U.S. comparisons, 138
  mammals, large, 15–16, 41, 196–97
  species of animals, attitudes toward different, 100, 102
  urban centers, 193
  zoos, 87
Affective learning, 211–12
Africa, 197
African Americans, 60–62
Age and American values toward nature/living diversity, 46–51
Aggression expressed through animal abuse, 97
Agriculture, 11, 121, 126, 186–90
Agriculture, U.S. Dept. of, 174
Alcohol and drug abuse, 96
Alkaloids, 11
Ambivalent nature of zoos, 85, 86, 89
American burying beetle, 124
American values toward nature/living diversity, 37, 63
  age, 46–51
  demographics on, 41–46

American values toward nature/living diversity (*continued*)
  education and income, 54–56
  ethnicity, 60–62
  gender, 51–54
  Japanese values compared to, 135–42
  methods used in finding, 38–41
  shifts in, 5
  species of animals, attitudes toward different, 101–2
  urban/rural variations, 56–60
  utilitarian value of nature, 10
  water projects that endanger fish species, 165–66
  whales, 118, 119
Anesthetized experience of film/television portrayals of nature, 90
Animal archetypes, 19
Anthropocentric view of nature, 133, 216
Anthropomorphism, 18–19, 91
Antihunting sentiments, 66, 69–71, 75–77
Antiwhaling sentiments, 119
Antiwildlife trade sentiments, 200
Ants, 128
Aquariums, 88–89
Arthropods, 25, 121
Artificial substitutes for nature, 15, 19
Asia, 123, 142
Assessing values, 39–40
Attitude questions, 39
Audubon, John J., 67–68, 80
Audubon Societies, 80–81
Augustine, Saint, 23
Auk, 33
Australia, 118, 119
Authoritarian decision making in wildlife management agencies, 204
Avarice and killing of whales, 111, 112, 114, 120
Avoidance of threatening aspects of nature, 25

Bacon, Roger, 23
Bald Eagle Protection Act, 157
Baleen whales, 121
Bantu people (Botswana), 147
Bats, 100
Bears, 101, 110
Beavers, 112
Bees, 125, 126, 128, 129
Beetles, 120, 123, 124, 128
Belief orientations, 212
Berry, Joyce, 52
Bettelheim, Bruno, 19
Biocides, 187
Biological assessments, 39
Biological character of the human species, 6
Biological diversity, 124, 204–7
Biological origin of human attitudes toward animal species, 102
Biomass of earthworms and arthropods, 121
Biophilia, 6–7, 26–34
Biophysical forces and conservation policy, 156, 207
Biota, abundant/diverse/healthy, 188–89
Birds:
  commercial hunting, 80–81
  demographics on, 120
  endangered species, 29, 196
  Hawaii, 30
  palila, the, 166–71
  perceptions of, 100, 101
  pigeons, passenger, 33–34, 67–68
  shorebirds, 157
  watching, 79–83
Bison, 157
Biting invertebrates, 102, 127
Black-footed ferrets, 171–76
Blue whale unit, 113
Bonding/kinship/caring value of nature, 21–22, 33
  see also Humanistic value of nature
Boston strangler, 94
Botswana, 146–49, 199
Bourjailly, Vance, 74
Breer, Paul, 77

Brewer, Garry, 157
Buddhist-Hindu perspective on nature, 132–42
Burger, Warren, 164
Bushmen (Botswana), 148, 149–52
Butterflies, 100, 124, 125, 129, 196

Cactus, 196
Caimans, 100
Callicott, Baird, 142
Campbell, Joseph, 19
Canada, 116–20, 119
Cancer, 11, 122–23
Caretaker role, 52
Caring/bonding/kinship value of nature, 21–22, 33
  see also Humanistic value of nature
Carnivores, 29, 66–67
Carter, Jimmy, 164
Caterpillars, 123
Cats, 22, 97, 101
Cattlemen and wolf killing, 104–5
Charismatic megavertebrates, 17
Cherokee people, 162
Children, 18, 19, 46–47, 93–98
China, 135
Clark, Tim, 155
Clarke, C.H.D., 72
Classifying/measuring people's values of wildlife/nature, 5–6
Cleaver, Eldredge, 61
Climax phases of mature forests, 187
Coastal whaling, 112
Cobras, 100
Cockroaches, 124, 127
Co-existence with nature, 134–35, 150, 217–18
Cognition and symbolic value of nature, 18
Cognitive learning, 211–12
College education, 54–56, 107
Colonial America and hunting, 67
Commercial hunting, 77–79
  birds, 80–81
  Canadians, 118
  elephants, 196–200
  Fish and Wildlife Service, U.S., 4
  international wildlife trade, 195, 201
  pigeons, passenger, 67–68
  support for, 119–20
  whales, 113, 114
Communication and symbolic value of nature, 19
Community involvement and conservation policy, 159, 178–80, 204
Competition and global loss of biological biodiversity, 186–94
Competition and self-worth, 52
Complexity of the wildlife policy process, 207
Concrete gratifications, 47
Conduct, logical/rational/abstract rules of, 52–53
Congress, U.S., 164, 199
Connections between people and the natural landscape, 189, 192, 211, 217
Consciousness sharing between humans and animals, 150
Conservation of biological diversity, 185
  administration of wildlife management, 203–8
  aesthetic value of nature, 170–71
  basis of, 27–28
  elephants, 196–200
  film/television portrayals of nature, 91
  forestry and agriculture, 186–90
  Japanese-U.S. comparisons, 140
  multidimensional task of, 155–56
  urban centers, 190–94
  utilization, wildlife, 194–96, 201
  wolf as a symbol of, 110–11
  zoos, 89
  see also Endangered species
Consumption patterns, 30
Consumptive activities, 65–79
Controlling animals, urge for, 96
Convention on International Trade in Endangered Species of Flora and Fauna (CITES), 5, 199
Cook, Captain, 30, 166–67

Cooperation/partnership and conservation policy, 180
Cooptation of wildlife, human, 186
Coral reefs, 121–22
Corn, 11
Cost-benefit analysis and conservation policy, 163
Coyotes, 101, 104
Crickets, 124
Crocodiles, 100, 196
Cruelty to animals, 92–98
Cultivating the ideal in nature, 139–40
Cultural relativism and animal abuse, 93
Culture influencing values of nature, 131
  Eastern and Western perspectives, 132–42
  Germany, 142–45
  hunter-gatherers, 149–52
  nonindustrial societies, 146–49
  species, attitudes toward different, 103

Darter species, 123, 161–66
Darwin, Charles, 23, 30, 167
Death in nature, ubiquity of, 77
Decomposition, 31, 120, 122
Defending biological diversity, wildlife management, 207–8
Deferred adolescence, 54
Deforestation, 11
DeLeon, Robert, 157
Demographics on American values toward nature/living diversity, 41–46
Demographics on animal species, 21
Denning, 53
Dependence on nature and living diversity, 7, 206, 217
Destruction of life on earth, 29–32, 82–83
  agriculture, modern, 187–88
  Asia, 142
  Bushmen, 151
  connections between people and the natural landscape, 217
  habitat destruction, 29, 135, 161, 168–69, 186, 198
  wolf killing, 106
Dialectical stress, 20–21
Diamonds, 147
Dinesen, Isak, 84–85
Dissociation from nature, 27, 193
Distancing and respect for nature, healthy, 25
Dogs, 22, 101
Dolphins, 78, 112, 115
Domestic animals, 21–22, 41–42, 52, 93, 101–2
Dominionistic value of nature, 6, 20–21, 32–33
  belief orientations, 212
  defining, 38
  demographics on American, 43–44
  hunting, 73–74
  invertebrates, 128
  Japanese-U.S. comparisons, 138
  science, 134
  whales, 111, 114, 116, 118
  wolves, 103–4
  zoos, 87, 89
Dread and dislike, sentiments of, 25
Drosophila (fruit fly), 123
Drug and alcohol abuse, 96
Duality separating people from nature, 133
Dubos, René 27
Dynamic wildlife policy, 207

Eagles, 100
Earthworms, 121
Eastern Pacific Gray whale, 112
Eastern perspective on nature, 132–42
Ecologistic/scientific value of nature, 6, 13–14
  attitudes toward animals influenced by, 99–100

birding, 81–82
black-footed ferrets, 175
cognitive learning, 212
defining, 38
demographics on American, 44–45
endangered species, 177–78, 182
ethnicity, 60
film/television portrayals of nature, 90–91, 92
hunting, 78
invertebrates, 120, 123–24
Japanese-U.S. comparisons, 138, 140, 141
management, wildlife, 204–5
marine parks, 88
urban/rural variations, 57
whales, 114–15
wolves, 108–10
zoos, 87, 89
Economic assessment, 39–40
Economic development vs. conservation policy, 163
Ecosystems, healthy, 120, 188–89, 201, 205, 207–8
Edible organisms, 121
Education:
American values toward nature/living diversity, 54–56
Botswana, 149
endangered species, 178, 184
invertebrates, 126
symbolic value of nature, 18
values/perceptions of wildlife/conservation, 210–13
whales, 118, 119
wolves, 107, 110
zoos, 86
Egypt, 123
Ehrenfeld, David, 112
Einstein, Albert, 23
Elderly people, 50–51, 107, 110, 118, 126
Elephant Protection Act of 1988, 199
Elephants, 196–200
Emotional attachment to animals, *see* Humanistic value of nature
Empiricism and conservation policy, 177
Endangered species, 5
black-footed ferrets, 171–76
causes of, three major, 29–31
conclusions about, 181–84
international wildlife trade, 196
intrinsic right to exist, 215
invertebrates, 124
Japan-U.S. comparisons, 136
lessons learned about protecting/recovering, 176–81
multidimensional task of conserving, 155–56
palila, the, 166–71
pigeons, passenger, 67–68
restoration efforts for, 107, 110, 172–75, 205
snail darter, 161–66
whales, 111–14
wolves, 103–5
Endangered Species Act of 1973 (ESA), 4, 106, 115, 157–60, 164, 174, 179
Energy and nutrient transfer, 120
England, 118, 119, 136
Environmentalists, 126, 173, 175, 176, 179
Environmental literacy/education, 210–13
Environmental protection movement, 81, 135
Ethic of care/compassion for the diversity of life, 213–18
see also Moralistic value of nature
Ethnicity and American values toward nature/living diversity, 60–62
Europe, 80, 85, 103
Excreta decomposed by invertebrates, 122
Experience/learning influencing affiliations with natural world, 26–27

Exploiting wild living resources, *see* Utilitarian value of nature
Extended family, 69
Extinctions, species, 7
see also Endangered species

Fairy tales, 19
Falcons, 196
Families, abusive, 96–98
Farmers, 57, 108–9, 110, 124, 126
Fearing animals, *see* Hating/fearing animals
Felthous, Alan, 92, 94
Ferrets, black-footed, 171–76
Film and television, portraying wildlife/nature on, 90–93, 141
Financial dependence on taxing/licensing sportsmen, wildlife management's, 207–8
Fish:
abusive behavior toward, 93
commercial fishers, 78, 118
endangered species, 196
Fish and Wildlife Service, U.S., 4
freshwater, 29, 161
popular species, 101
recreational fishers, 69
snail darter, 161–66
Fish and Wildlife Service, U.S. (FWS), 4–6, 40
Five year olds and under, 47–48
Fly, fruit, 123
Food webs, 31
Forestry, 121, 186–90
Formal education, 212
Four Stages of Cruelty (Hogarth), 93
Free Willy, 88
Freshwater fish, 29, 161
Fur trapping, 78, 108–9

Galápagos Islands, 30
Game animals, 102, 169–70
Garden insects, 124
Gender and American values toward nature/living diversity, 51–54, 74
General Accounting Office (GAO), 165–66
Genetic/biochemical/physical properties of plants/animals, 10–11, 22–23, 84
Geographical/ecological variation and biodiversity characteristics, 136–37
Germany:
birding, 80
film/television portraying nature, 90
species, attitudes toward different, 102
values/perceptions of wildlife/conservation, 142–45
whaling, 118, 119
zoos, 85
Gila National Forest of New Mexico, 106
Gilligan, Carol, 52
Giraffes, 84–85
God Squad, 164
Goldman, William, 105
Government and conservation policy, 173–74, 179–81, 202–3
Graphard, Alan, 140
Greece, 85
Grizzly bears, 110
Guns as symbols of maleness, 74

Habitat destruction:
competition, 186
elephants, 198
freshwater fish species, 161
Japan and China, 135
palila, the, 168–69
reasons behind, 29
Hardin, Garrett, 113
Harvesting practices, 53, 135
Hating/fearing animals:
invertebrates, 126, 129
prairie dogs, 171, 174–75

species or breed, particular, 96–97
wolves, 25–26, 104, 110
Hawaiian Islands, 30, 166–71
Hawaii Department of Land and Natural Resources (HDLNR), 167
Health/resilience of ecosystems influenced by invertebrates, 120–21
Health standards, public, 188
Heerwagen, Judith, 17
Hierarchical structures in wildlife management agencies, 204
Hierarchical ties, 53
Higher vertebrates and animal abuse, 93
Hillman, James, 128
History and culture influencing attitudes toward different species, 103
Hogarth, William, 93
Holsinger's groundwater planarian, 124
Honey, 122
Honeycreepers (birds), 30, 166–71
Hornaday, William, 105
Horses, 101
Human benefits from ecological functions of invertebrates, 122
Humanistic value of nature, 6, 21–22, 32
abusing animals, 93
affective learning, 212
Botswana, 148–49
defining, 38
demographics on American, 41–42
education, 211
elephants, 197
film/television portrayals of nature, 90
gender, 52, 53
hunting, 74–75, 78
invertebrates, 124, 125
Japanese-U.S. comparisons, 136
species of animals, attitudes toward different, 102
urban/rural variations, 59
whales, 114, 118
zoos, 87
Human management, 3–4
Human uniqueness, 133
Humpback whales, 115
Hunter-gatherers, 149–52
Hunting, 65
antihunting sentiments, 69–71, 75–76
carnivore/vegetarian debate, 66–67
endangered species, 67–68
Fish and Wildlife Service, U.S., 4–5
meat, 72–73
nature, 70–72
revenues from licenses/taxes, 4, 167, 207–8
sport, 4–5, 68–69, 73–75
wolves, 108–9
see also Commercial hunting

Iceland, 115
Identity/fulfillment dependent on expressing values of nature, 9–10, 27
Identity of humans threatened by numbers of invertebrates, 128
Illness linked to invertebrates, 127
Iltis, Hugh, 28
India, 14
Indigenous peoples in Third World countries, 146–49, 200
Indirect encounters with animals, 65, 83–89
Individual liberties and conservation policy, 159
Industrial-agricultural society, 126
Inexhaustibility fallacy and whale killing, 113–14
Informal education, 212–13
Infrastructural support for birding, 83
Innate tendency to focus on life and lifelike processes, 6–7
Insects:
abusive behavior toward, 93
attitudes toward, 126
demographics on, 22, 120
ecologistic/scientific value of nature, 13, 45, 124
endangered species, 29

Insects (*continued*)
  keystone species, 121
  music inspired by, 123
Institutional-regulatory dimension and conservation policy, 156
Integrated wildlife management, 207–8
Integrity/harmony/balance in nature, striving after, 16–17
Intellectual stimulation/satisfaction, 13, 14, 18
Intergovernmental competition and confusion, 179
Interior, U.S. Dept. of, 174
International wildlife trade, *see* Commercial hunting
Interviews, personal, 40
Intrinsic right to exist, 215
Invertebrates, 13, 100, 120–29, 196
Ireland, 136
Isopods, 124
Ivory, elephant, 196, 198–99

Japan:
  birding, 80
  film/television portraying nature, 90
  invertebrates, 102
  ivory trade, 198
  love of nature, 134–35
  utilitarian value of nature, 10
  views of nature, contemporary, 135–42
  whaling, 115, 118
Jesus, 23
Judeo-Christian religious beliefs, 133
Jung, Carl, 19
Jurisdictional authority and whale killing, 113

Kalahari Desert, 146
Kauai cave wolf spider, 124
Kenya, 79
Keys scaly cricket, 124
Keystone species, 121, 183, 196–97
Killer whales, 115
Kinship/caring/bonding value of nature, 21–22, 33
  see also Humanistic value of nature
Knowledge of nature, *see* Ecologistic/scientific value of nature
Kohlberg, Lawrence, 46, 47, 52
Kozicky, Ed, 72
Krill, 121
Krutch, Joseph W., 76, 184

Lacey Act of 1900, 81, 157
Ladybugs, 100, 125
Land Management, Bureau of, 174
Land ownership, 57, 59, 159, 173, 178–79
Landscapes and aesthetically salient animals, 16
Language and symbolic value of nature, 16–17, 20
Large animals, aesthetic emphasis on, 15–16, 41, 196–97
Latitudinal range and biodiversity characteristics, 136
Lawrence, Elizabeth, 18
Learning/experience influencing affiliations with natural world, 26–27
Leghold trap, 70
Legislation, environmental, 5, 81, 157, 188
Leopold, Aldo, 3, 13, 16, 28, 33, 70, 105–6, 192
Leopold, Starker, 187
Lévi-Strauss, Claude, 19
Liberties and conservation policy, individual, 159
Licensing/taxing sport hunters, 4, 167, 207–8
Limited-species agricultural/forestry practices, 187
Little Tennessee River, 161–63
Livestock producers, 107, 173, 174–75
Living diversity, 32

Local rights/support and conservation policy, 159, 178–80, 204
Lopez, Barry, 25, 104, 111
Louis XIV, 85

Madagascar, 11
Madson, John, 72
Mamani-naio forest (Hawaii), 167, 168
Mammals:
  aesthetic response to large, 15–16, 41, 196–97
  demographics on, 120
  endangered species, 29
  Hawaii, 167
  Japan, 136
  marine, 88–89, 112
  species, attitudes toward different, 101, 102
Management, human, 3–4
Management, wildlife, 3
  administration, 203–8
  Botswana, 148
  debate on, contentious, 42–43
  Eastern attitudes, 142
  elephants, 199, 201
  Endangered Species Act of 1973, 158–59
  Fish and Wildlife Service, U.S., 4–5
  game species, emphasis on, 169–70
  gender, 53
  Germany, 143, 145
  reform of wildlife institutions, 184
  sport hunters, 68–69
  urban/rural variations, 59–60
  utilization, 194
  whales, 115–16
Manatees, 112
Manure decomposed by invertebrates, 122
Marginalized peoples and hunting, 67
Marine Mammal Protection Act of 1972, 4, 115, 116
Marine mammals, 88–89, 112
Marine parks, 88–89
Market hunting, *see* Commercial hunting
Masculinity and sport hunting, 74
Matthiessen, Peter, 33, 113
Mature forests, 187
Mauna Kea, 167
Maximum sustainable yield, 201
Meaning and aesthetic attraction, 17
Meat hunting, 72–73
Media, mass, 213
Medicines and invertebrates, 122
Medicines and plants, 11
Megavertebrates, charismatic, 17
Meiji period (Japan), 141
Men and American values toward nature/living diversity, 51–54
Mental competence sharpened through dominionistic value of nature, 20–21
Mental health preserved through the cult of the wilderness, 27–28
Metaphoric expression, 18–19
Mexico, 11
Mice, 97
Michigan study on wolves, 107–9, 175
Microbial organisms, 13, 122
Mimicking natural processes, 14
Mindlessness/madness associated with invertebrates, 128
Minnesota study on wolves, 107–9, 175
Modern whaling era, 112
Mollusks, 161
Monocultures, agricultural/forestry, 121, 126, 187, 189
Montana, 175, 176
Moralistic value of nature, 6, 22–24
  belief orientations, 212
  birding, 81
  black-footed ferrets, 175
  defining, 38
  demographics on American, 41, 42–43
  gender, 52–53
  hunting, 75–77
  invertebrates, 125
  Japanese-U.S. comparisons, 138, 140, 141

Moralistic value of nature (*continued*)
urban/rural variations, 57
whales, 114, 118
Moratorium on killing whales, global, 114, 115, 157
Mordi, Richard, 146, 147, 149
Mountain lions, 110
Multidimensional task of conserving endangered species, 155–56
Musical composition inspired by invertebrates, 123
Myths, 19

National Environmental Policy Act of 1969, 5
Nation-state, modern, 152
Native Americans, 103
Naturalistic exhibits in zoos, 86–88
Naturalistic value of nature, 6, 11–13, 32–33
black-footed ferrets, 175
cognitive learning, 212
defining, 38
demographics on American, 44
ethnicity, 60
film/television portrayals of nature, 91
hunting, 78
invertebrates, 125
Japanese-U.S. comparisons, 136, 139–40
urban centers, 191–93
Nature hunting, 70–72
Negativistic value of nature, 6, 24–26
affective learning, 212
black-footed ferrets, 175
Botswana, 149
defining, 38
demographics on American, 41, 42
ethnicity, 60
Japanese-U.S. comparisons, 136
whales, 114
Nelson, Arvid, 143
Nelson, Richard, 150, 151, 217
Never Cry Wolf, 91
New England whaling industry, 112
Newfoundland, 118
New Haven (Connecticut), 191
Nickjack cave isopod, 124
Nonconsumptive uses of nature, 4–5, 55, 79–83, 194–95
Nonindustrial societies, 10, 146–49
Nonnative species causing destruction, 30, 167–68, 169
Norris, Kenneth, 112
Northern Fur Seal Treaty, 157
North Platte montane butterfly, 124
Norway, 115, 118
Numbers of invertebrates threatening human identity, 128
Numeration and conservation policy, 177
Nutrient and energy cycling, 31

Octopus, 123
Okavango swamps (Botswana), 146
Oklawaha sponge, 124
Orchids, 196
Organizational behavior and conservation policy, 180–81, 202–3
Orians, Gordon, 17
Origins of animals, common, 21–24
Ortega y Gasset, José, 66–67, 71, 72, 77
Oxygen production, 31

Pagan animism, 133
Palila, the, 166–71
Palila v. Hawaii Department of Land and Natural Resources, 168
Paradoxes of modern life, 66–67
Parasitism, 120, 121, 127
Parks, national, 79
Parrots, 196
Participatory/team-oriented wildlife management agencies, 203–4
Periwinkle, rosy, 11
Personal utilitarian experience of nature and living diversity, 11
see also Naturalistic value of nature

Pest control and invertebrates, 122
Pets, 21–22, 41–42, 52, 93, 101–2
Pharmaceutical products, 11
Physical/behavioral characteristics and attitudes toward animals, 99, 102
Physical competence sharpened through dominionistic value of nature, 20–21
Physical fitness, 13
Piaget, Jean, 46–47, 52
Pigeons, passenger, 33–34, 67–68
Planarians, 124
Planning, long-term strategic, 205
Plants, 11, 31, 120–22, 196
Platter, Zygmunt, 164
Pleasure from connection to nature and living diversity, *see* Naturalistic value of nature
Poison control techniques, 53
Political/economic interests and conservation policy, 160, 163–64, 168, 178, 182
Pollination, plant, 31, 120–22
Population size/distribution and biodiversity characteristics, 136
Positional ties, 53
Power and authority, distribution of, 53
Prairie dogs, 171–76
Predators, 25, 29, 101, 110–11, 120, 136
Prejudice against a species or breed, *see* Hating/fearing animals
Preventative conservation policy, 182–83
Property rights and conservation policy, 159, 169, 173, 178–79
Protection and aesthetic value of nature, 15–16, 17, 41, 100, 102, 196–97
Protection, ethic of species, 215–16
Protection movement, environmental, 81, 135
Protections for whales, international, 114, 115, 157
Psychological and behavioral consequences of abusing animals, 65, 92–98
Psychological stress and captive marine animals, 88
Public broadcasting, 91
Public land access and conservation policy, 173, 175, 178–79
Public understanding/appreciation of living diversity, 183
Purposeful design and aesthetic attraction, 17

Quality of life and healthy ecosystems, 188–89, 191–92

Ranchers, 173, 174–75
Rangelands, public, 173, 175, 178–79
Rats, 93, 97, 100, 104
Rattlesnakes, 93, 100, 104
Reading, Richard, 174, 175
Recreation, 12, 52, 70, 137
  see also Hunting
Reform of wildlife institutions, 184
Regulations, government, 173–74
Relativism and animal abuse, taxonomic/cultural and, 93
Reptiles, 196
Research:
  abusing animals, 92–98
  American values toward nature/living diversity, 38–41
  artificial substitutes for nature, 15
  avoidance of threatening aspects of nature, 25
  black-footed ferrets, 175–76
  film/television portrayals of nature, 90
  Fish and Wildlife Service, U.S., 4–6
  invertebrates, 123–26
  Japanese and American views of nature, contemporary, 135–42
  naturalistic value of nature, 12
  species of animals, attitudes toward different, 101–2
  whales, 116–20
  wolves, attitudes toward, 107–10
  zoos, 84, 86–87

Resource-dependent populations, 57, 59
Resource Policy Framework, 156–57
Respect for nature, 25, 201
Restoration efforts for endangered species, 107, 110, 172–75, 205
Retaliating against an animal, 96, 97
Revenue from taxing/licensing sport hunters, 4, 167, 207–8
Rhinoceros, 196
Right whales, 112
River systems, alteration of, 161
Robins, 100
Rolston, Holmes, 17, 20
Romantic appreciation of the natural world, 144
  see also Humanistic value of nature
Rome, 85
Roosevelt, Theodore, 105
Rural areas:
  endangered species, 178
  meat hunting, 72–73
  population decline, 69
  urban areas compared to, 56–60
  whales, 118
  wolves, 107–9

Sacrificing human needs for rights of animals, 144
Sadism and animal abuse, 97–98
Saito, 140
Salmon, 100
Salt River (Arizona), 30
Scarab beetles, 123
Schaller, George, 16
Schulz, Wolfgang, 142, 143
Science controlling nature, 134
Scientific understanding of nature, *see* Ecologistic/scientific value of nature
Scientific value of invertebrates, 123, 126
Seals, 112, 118
Seed dispersal, 120, 121
Seilstad, G., 12
Self-centered thinking, 47
Self-interest, expanded concept of personal/social, 214–15
Self-worth, 52–53
Sensitivity to detail, 12
Serpell, James, 22
Sharks, 100
Shellfish, 122, 125, 196
Shepard, Paul, 18, 70
Shocking people for amusement using animal abuse, 97
Shorebirds, 157
Shrimp, 125
Sierra Club, 168
Similarity of species to human beings, biological, 102, 197–98
Single-species agricultural/forestry practices, 121, 126, 187, 189
Snail darter, 123, 161–66
Snakes, 25, 93, 97, 100, 104
Social diversity, managing for, 204, 206–7
Social-psychological techniques, 39, 40
Social surveys, 38
Social ties strengthened through zoos, 86
Sociobiology, 123
Socioeconomic factors in conservation policy, 159, 163, 173, 178, 182
Socioeconomic status, 28, 54–56, 60
Soil loss/compaction, 187
Songbirds, 100
South Africa, 199
South America, 123
Species of animals, attitudes toward different:
  American values toward nature/living diversity, 101–2
  factors influencing, four major, 99–100
  game animals, 82, 169–70
  hunting/fearing animals, 96–97
  interconnections, 205
  invertebrates, 120–29

mutualism and competition, 120
protection ethic, 215–16
whales, 111–20
wolves, 103–11
Spiders, 25, 93, 120, 123, 124, 127, 128
Spiritual kinship between humans and animals, 24, 123, 150–51
Splitrail value of hunting, 70
Sponges, 124
Sport hunting, 4–5, 68–69, 73–75, 167, 207–8
State rights and conservation policy, 159
Steinbeck, John, 23
Stinging invertebrates, 102, 127
Stories and symbolic value of nature, children's, 19
Stress and dominionistic value of nature, 20–21
Stress mitigation, 12
Subjugation of nature, 133
Substitutablity fallacy and whale killing, 113–14
Suburbanization of the American countryside, 56–57
Supreme Court, U.S., 164
Surveys, 38
Survival prospects of the human species, 66
Survival skills sharpened through dominionistic value of nature, 20–21
Sustenance and aesthetic value of nature, 17
Swans, 100
Swensen, Susan, 86
Symbolic value of nature, 6, 17–20, 32, 38, 212

Taiwan, 198
Taxing/licensing sport hunters, 4, 167, 207–8
Taxonomic relativism and animal abuse, 93
Team-oriented/participatory wildlife agencies, 203–4
Technology, 113, 134, 188
Teenagers, 45–46, 49–50
Tellico River dam, 161–64
Tennessee Valley Authority (TVA), 161–63
Termites, 124
Theistic value of nature, 148, 149
Therapy, animal assisted, 21
Theriophobia, 26
Third World countries, 146–49, 200
Thomas, Keith, 21, 93
Tigers, 196
Timber production, large-scale, 186
Tourism, nature, 79, 83
Toxicants, 187
Transience and mobility of Americans, 69
Transitioning from disconnection to unity and purpose, xvi–xix
Trappers, 78, 108–9
Tribal peoples, 23, 149–52
Tropics, 120, 121
Trout, 100
Tuna, 78, 196
Turtles, sea, 196

Ubiquity of death in nature, 77
Ubiquity of invertebrates, 128–29
Ulrich, Roger, 12, 15, 25
Ungulates, grazing, 100
United States:
birds popular in, 80
film/television portrayals of nature, 90
invertebrates, 124
transcience and mobility in, 69
200 mile zone of economic control along coasts of, 116
whales, 116
wolves, 103–4
zoos, 85
see also American values toward nature/living diversity

Unity of all things, 17, 22–24, 134
Universal character of aesthetic responses to nature, 15
Urban centers:
  biophilia hypothesis, 28
  conservation of biological diversity, 190–94
  Montana, 176
  rural areas compared to, 56–60
  whales, 119
  wolves, 107, 110
  zoos, 84
Utilitarian value of nature, 6, 10–11, 14
  belief orientations, 212
  black-footed ferrets, 175
  Botswana, 148, 149
  conservation of biological diversity, 194–96, 201
  defining, 38
  demographics on American, 41, 42–43
  education, 54
  elephants, 196–200
  ethnicity, 60
  gender, 51
  Germany, 144
  hunter-gatherers, 151
  hunting, 72
  invertebrates, 120
  Japanese-U.S. comparisons, 136
  species protection, ethic of, 216
  urban/rural variations, 57
  whales, 113, 114, 116, 118

Valuational dimension in conservation policy, 156
Valuational sustainability, 207
Values/perceptions of wildlife/conservation, 3, 99
  aesthetic, 14–17
  biophilia, 26–34
  conservation of biological diversity, 188–94
  dominionistic, 20–21
  ecologistic/scientific, 13–14
  education, 210–13
  endangered species, 176–77
  ethic of care and compassion, 213–18
  Fish and Wildlife Service, U.S., 4–6
  Germany, 142–45
  humanistic, 21–22
  identity/fulfillment dependent on expressing, 9–10
  management, wildlife, 206–7
  moralistic, 22–24
  naturalistic, 11–13
  negativistic, 24–26
  symbolic, 17–20
  typology of basic, 38
  utilitarian, 10–11
  see also American values toward nature/living diversity; *individual value headings*
Vegetarians, 66–67
Vertebrates, 93
  see also Mammals
Vicarious experience of nature, 65, 91–93
Violence in adulthood associated with childhood animal abuse, 93–98
Vitality/awareness, naturalistic experience and, 12
Volcanoes National Park, Hawaii, 167
Vultures, 100

Walrus, 112
Waste decomposition, 31
Watanabe, Hiroyuki, 134–35, 141
Watching wild animals, 79–83
Water production, 31
Water projects that endanger fish species, 165–66
Western perspective on nature, 93, 132–42
Westervelt, Miriam, 47
Wetlands, 146–47
Whales, 78, 79, 111–20, 157
White, Lynn, 133
Wilson, Alexander, 80

Wilson, Edward O., 6, 24, 26–27, 29, 192, 215, 217–18
Wolves, 25–26, 101, 103–11
Women and American values toward nature/living diversity, 51–54, 126
Work, 52
Worms, 121, 124
Wyoming and black-footed ferrets, 172

Yellowstone National Park, 107, 172
Young, Stanley, 104
Young adults, 50

Zimbabwe, 199
Zoos, 83–89, 141, 213